303.483 LEM 2005

Lemov, Rebecca M. (Rebecca
World as laboratory :
experiments with mice,

This book is presented to

L.W. Nixon Library

in memory of

Linzie Williams

father of Kent Williams

by Butler Community College

May 30 , 2006

WORLD AS LABORATORY

World
as
Laboratory

EXPERIMENTS WITH

MICE, MAZES, AND MEN

REBECCA LEMOV

HILL AND WANG

A division of Farrar, Straus and Giroux
New York

Hill and Wang
A division of Farrar, Straus and Giroux
19 Union Square West, New York 10003

Copyright © 2005 by Rebecca Lemov
Distributed in Canada by Douglas & McIntyre Ltd.
Printed in the United States of America
First edition, 2005

Library of Congress Cataloging-in-Publication Data
Lemov, Rebecca M. (Rebecca Maura)
    World as laboratory : experiments with mice, mazes, and men / Rebecca Lemov.— 1st ed.
        p.   cm.
    Includes index.
    ISBN-13: 978-0-8090-7464-8 (hardcover : alk. paper)
    ISBN-10: 0-8090-7464-8 (hardcover : alk. paper)
      1. Social engineering—United States—History—20th century.   2. Behavior modification—
Study and teaching—United States.   3. Psychoanalysis.   I. Title.

    HM668.L46 2005
    303.48'3'0973—dc22

                                                                              2005006882

Designed by Patrice Sheridan

www.fsgbooks.com

10 9 8 7 6 5 4 3 2 1

TO PALO AND IVY

CONTENTS

WORLD AS LABORATORY

THIS BOOK IS ABOUT AN IDEA that lived and died, then lived again during the key years of an undertaking often known as "the American Experiment." The idea was quite simple: If one could quantify and control the internal arena of the personal self—its urges and wants, its worries and fears—then the running of a modern society would require less brute external force. In the long term, putting this idea into practice would make it possible to regulate human beings in tune with the needs, demands, desires, and models of the social order, so that people would want to do whatever they were instructed to do (for example, to die for one cause, shop for another; lie back in some instances or rise up in others). This was human engineering, an endeavor also known at times as behavioral engineering, social engineering, or environmental engineering.

Starting around the turn of the twentieth century in laboratories across America, scientists acted in pursuit of the idea: they concentrated on the most human part of the equation, leaving no room for what sentimentalists would call the soul, and attempted to bring the conditioning process to bear on all aspects of life. They turned the idea into an activity, a program, and eventually something akin to a system. The social scientists, foundation officers, and policymakers who made up the ranks of human engineers might have preferred a different label—perhaps "pragmatist," "experimentalist," "behaviorist," "progressive," or even plain "liberal"—but still they shared a sense that one could devise ways to predict and control people's actions and behaviors—as well as, eventually, their thoughts.

Of course, the general idea of a science of human behavior had bewitched people for a long time. Europe in the seventeenth, eighteenth, and nineteenth centuries did not lack for philosophers as well as eccentrics who fixed on certain propositions: perhaps there was an underlying mechanism to the body and soul, perhaps thought itself was susceptible to alteration. To take a key example, La Mettrie's strange treatise *Man a Machine*, published in 1748, argued that the universe contained nothing but matter and motion. Before this came Hobbes's *Leviathan*, with its "Artificial Animal" to be made, and Descartes's mechanistic views of the body's hydraulic function; afterward came Comte's and Saint-Simon's dreams of a true science of society. Accompanying these writings were a parade of actual devices: humanlike machines and machinelike humans; mechanical eagles, mechanized servants, and toys that came to life; automatons that inhabited gardens, courts, and rooms to carry out the will of their builders; and machines that simulated living things but were completely controlled. Yet these efforts did not bring together theory and practice. The people who dealt in ideas were hardly ever the ones who built the devices.

During the first two-thirds of the twentieth century, theory and practice met in certain laboratories in America, as human engineering began to take the reality-driven form of a series of experiments. Adopting what the proto-biotechnologist Jacques Loeb called an "engineering standpoint" toward human life, its adherents set about to make the ultimate social science. In the 1920s the visionary technocrat Beardsley Ruml defined exactly what this new science would be and funneled great amounts of Rockefeller Foundation money into it, the equivalent of several billion of today's dollars—a far larger amount than the federal government could have come up with at the time. The result—a combination of psychology, sociology, anthropology, and psychoanalysis, with contributing subdisciplines of economics, political science, theology, mathematics, physiology, ethology, and ecology—was a social science not previously encountered in human history, one that could only have emerged out of the peculiar life-size petri dish where American ambition, open space, and can-do approach combined in the agar of scientific advance.

THIS BOOK TELLS A SECRET HISTORY that is not really secret. The obscurity attaching to the events here recounted is mainly the result of their having been ignored for some time, and the story of human engineering is in fact available to anyone with a library card. During the heyday of this emerging science of human behavior and society, experts installed themselves in laboratories and availed themselves of animal and human subjects, constructing ever-finer measuring devices and recording apparatuses. They set about building tiny labyrinths, small-scale social situations, miniature restraining devices, and microcosmical dioramas of real life where test animals grappled in conditions simulating war, competition, or self-doubt—all within the four walls of a laboratory. They adapted or invented an array of tools, gadgets, techniques, and models in order to gather social and behavioral facts. Among them, by the 1920s and 1930s, were artificial logic machines, "problem boxes" and "special cages," mazes of every possible contour and provenance, simple hospital beds reserved not for sick but for "normal" people as material for observation, galvanic skin-response recorders (precursors to lie-detector machines), punishment grills for delivering electric shock, the "analytic situation" as a model for an obedient society, and a compendious anthropological filing cabinet—Remington Rand, grade A, olive drab, to be exact—that was capable, at least theoretically, of containing the sum of all important information from every culture in existence.

By the 1940s, with the world war as their engine, behavioral engineers made use of mainframe computing and cybernetics to invent new technologies of information collecting and data processing. By means of the largest stack of punch cards ever assembled, to take just one example, a great many U.S. Army soldiers during World War II were "processed" using Hollerith codes and data profiles, so that everything from their feelings about mess hall to their loyalty to their country could be measured, targeted, studied, and potentially changed. Morale and the desire to fight could be built from scratch. Postwar developments benefited from more advanced information technologies, and the machines in turn became more sophisticated. Banks of computers ran cards in batches, and IBM machines worked overtime into the night. One automaton acted like a lab rat, while real lab rats' actions were graphed and rendered as algorithms. At Harvard's Laboratory of Social Relations, a "special room" was equipped with a bevy of recording devices, flashing feedback lights, and one-way mirrors for graphing the minutiae of human encounters—the origin of the modern focus group. Also in the making were polling on a mass level; polygraphs targeting particular individuals; tests for registering motivation, intelligence, loyalty, and every manner of mental twitch; propaganda of black, white, and gray varieties; and programs for the large-scale gratification of proliferating desires. Other devices rendered the inner states of shoppers or voters as a series of numbers. Altogether these developments comprised a "new magic of electronics," as a group of Harvard professors hailed them, ready to bring about a "sci-fi future."[1]

Soon these devices, built up from years of research on laboratory animals, led to experiments on human subjects that combined drugs, psychosurgery, and other alarming manipulations. Seemingly straightforward programs in human engineering turned out to have many unanticipated applications in a high cold war–era atmosphere fraught with enemies capable of the direst maneuvers and unseemly psychological tricks. To combat these enemies, the Central Intelligence Agency, aware of the promises that social scientists had been making for some decades about behavioral and psycho-cultural engineering, employed social scientists in an array of projects designed to manipulate and control human behavior. Some were asked to work on interrogation techniques, setting conditions for the optimal retrieval of information; some trained soldiers to resist Soviet and Chinese Communist brainwashing; and others developed a remote-controlled Pavlovian cat that could spy on clandestine conversations or, with the push of a button, detonate. As the late 1950s became the early 1960s, the cartoonish, the conspiratorial, and the very real all came together.

WHAT GAVE HUMAN ENGINEERING its peculiar power as a social force—a "soft" power by and large, one that directed individual behavior less by using active force than by shaping the surrounding environment—was not simply its clever devices for measuring what had not before been measured. At the core of the movement resided a conviction, an idea, a philosophy, and finally a living reality: that the insights of Freud could be merged with the science of behaviorism.

The experimental trajectory reached its high point with a large-scale project to bring Freud's insights concerning the unconscious within the compass of human engineering. Psychoanalysis and behaviorism came together to make an all-purpose, American-style therapeutic-slash-engineering technique. Indeed, the great engine of psychoanalysis was harnessed for behaviorism deliberately and thoroughly, through the work of four experimentalists over five years during the late 1930s and early 1940s. This merger took place at a once eminent, now little-known concern at Yale University, the Institute of Human Relations. Here a properly "engineered" society with properly conditioned parts was said to hold the promise of replacing older forms of order (such as religion, tradition, and the dead hand of authority) with an infinitely more subtle form of social control.

Experiments in controlled spaces led to a core insight: that the process of observing and measuring reality as it unfolds within an experimental design will itself bring about all kinds of changes. This could be called the laboratory imagination, the power to generate choreographed changes—for good or for ill—from experiments run through clever apparatuses. What happened in laboratories was not less like reality, but more so.

The experiments here described—however justly they have been rel-
egated to the dustbins and graveyards of out-of-date scientific fashion,
however firmly scientists claim to have moved beyond their crude
excesses—could not be more intimately related to the way most Americans construct their living environments and sense of themselves in
the early twenty-first century. Human engineering, like other grand
ideas, triggered a cascade of consequences both intended and unin-
tended. This book traces the paths by which it left the laboratory and
entered the world.

PART ONE

Mazes:

Into the

Laboratory

CHAPTER 1

Strange Fruits and Virgin Births

ALMOST FIVE HUNDRED YEARS have passed since Francis Bacon imagined in *The New Atlantis* a scientific utopia whose inhabitants—in addition to controlling the wind and tides, turning salt water into fresh, and effecting the spontaneous generation of "frogs, flies, and divers others" out of thin air—perfected a method for turning plants and animals into new forms. Fruit trees were induced to bloom earlier or later, their fruit sweeter or different in size, smell, and shape from the typical. In the island's parks and enclosed pastures, all sorts of beasts and birds resided for purposes more experimental than ornamental: "We make them greater or taller than their kind," stated the great "Father" scientist in charge. "We find means to make commixtures and copulations of different kinds; which have produced many new kinds." Yet despite the marvelous specificity of his vision, Bacon did not offer a date for the actual start of these transformative practices. For when *The New Atlantis* was published in 1627, a year after Bacon's death, making one plant or animal turn into another was still the stuff of imaginary islands and magically endowed inhabitants.[1]

Nearly three hundred years later, coinciding with the turn of the twentieth century, two men in America were hard at work on precisely such projects, creating new plant and animal forms out of the old in, respectively, an experimental farm and a scientific laboratory. Over the course of his life Luther Burbank, the author of *Burbank's New Creations in Trees, Fruits and Flowers*, "built" between eight hundred and one thousand new hybrids, using a combination of mass production methods and great doses of patience: Idaho potatoes, raspberries-married-to-strawberries, and oversize walnut trees that could bear a ton of nuts in a single season all testified to his powers. The second man, Jacques Loeb, constructed a stable of creatures he called "durable machines" in his laboratory at the University of Chicago: two-headed marine worms, metamorphosed slime molds, hydras with mouth and anus reversed, and artificially propagated sea urchins. (For this last, he was known as the instigator of a new virgin birth in the lab and was nominated for the Nobel Prize.) Both men inspired awe in their day, one as a wizard, the other as a prophet, and both brought assembly-line-like methods to biological processes. But it is Loeb's self-proclaimed "technology of living substance" that provides an understanding of the birth of human engineering in America. For if it is common today to eat Burbank's fruits, it is just as common to live Loeb's ideas.

The German-born Loeb was not one for photo shoots; nor was he interested in larding the national dinner table or making farmer's lives easier. Unlikely as it may seem, his most influential work got its start in the great fertile plains of the midwestern United States. This locale was strangely appropriate: a Middle European scientist settled in Chicago and, with daylight factories, factory-style farms, and sleek silos all around him, went about revolutionizing the life processes themselves. Loeb was a new kind of visionary in whom ideas were not separate from activity.

Born in 1859 in Mayen, a small Rhineland town full of Catholics, Jacques Loeb was the first son of a fairly prosperous Jewish merchant-ing couple. At birth he was given the distinctly un-French appellation of Isaak, and for his first fifteen or so years, he faced the limitations of

life in the provinces; as an outsider because of his faith, he did little so-
cializing with other children aside from his brother. For the next ten
years, after moving to Berlin, he continued to face the double demerit
of provincial origins and Jewish blood. By the time he was a teenager,
his parents had died of illness, leaving him and his brother each a solid
though not spectacular living. At the behest of relatives, Loeb tried his
hand at a banking career in the metropolis but discovered the work to
be a "terrible bore," boredom seeming to him then and remaining for
the rest of his life his greatest enemy.[2] He resolved to diverge from the
path laid out for him by the *Realschule* (vocational) education that his
parents had preferred for their sons by enrolling in an elite school for
Jews.

Loeb proved to be quite brilliant at classical studies and to have an
ear for German literature, but he was impatient with the humanism,
theology, Hebrew, and philosophy that together comprised a well-
rounded education there and chose medicine instead. At twenty he entered medical school and renamed himself Jacques, thus setting for
himself a more Continental tone. At the time of his education in Germany, medical training was de rigueur for anyone who wanted to secure a professorship in the field of physiology. It was highly irregular to
train, as Loeb ended up doing, with agriculturalists and botanists as
well, and his decision to cross boundaries had something to do with
his marginal social status within the mandarin world of German academics. Forced to the sidelines, he had greater freedom in his training.
A disciplinary hybrid himself, he set the stage for the work he would
later carry out in midwestern America, constructing hybrid life-forms
out of stripped-down mechanical parts and functions.

DURING THE 1880s, M.D. in hand, Loeb went to work at the Berlin
Agricultural College under a professor known for his technique of
water-jetting away portions of a dog's brain. The point of the research
was to show that, once the shock of losing a portion of its brain had
worn off, the dog was less hampered than one might expect and, as it
often turned out, could still function quite well. Loeb's own work used

a similar method (making lesions on different areas of the brain instead of removing them) but went beyond his mentor's. He set out to find nothing less than an equivalent between physical and mental "energy." Convinced that thoughts and actions did not take place in separate spheres, he proposed a unified field theory for behavior. At the time, other scientists were wont to employ what Loeb felt were inadequate explanations for the behavior of their surgically altered laboratory animals. (For example, one of his colleagues argued for the existence of a "spinal soul," to be found diffused in the vertebral area.) Loeb, with a brashness not entirely attributable to his youth, was satisfied with none of these faith-based theories.

After suffering initial setbacks, a dog that had lesions made on its brain would soon be able to get on with its life and learn new ways of grooming itself, walking, or feeding. The whole organism, Loeb stressed, was a system of interconnected functions in dynamic equilibrium. On the basis of this observation, the postgraduate student set out to solve the age-old mind-body dilemma for the purposes of his dissertation. Loeb's analysis was hailed by none other than William James, the great American pragmatist, who commended him in the brain-function chapter of his *Principles of Psychology* for having "broader views than anyone."[3]

Somewhat hokily but in tune with the practices of the time, the young apprentice relied for dramatic effect on the performance of a dog whose motor centers governing its hind legs had been removed yet was still able to walk and beg. Notwithstanding such feats overcoming surgical setback, "dog work" was not to Loeb's liking, as it was messy and inexact. One didn't always know which part of the brain had been discommoded, nor which of the animal's difficulties might be attributable to infection, rather than surgery. When he presented his findings to the major German scientific congress of the day, dog included, a hostile colleague took hold of the demonstration animal and put him on the windowsill with his lower leg dangling into a flowerpot, thus counterdemonstrating that the dog was not in fact able to withdraw his leg. Humiliated, Loeb was rendered temporarily dumb. And so it was with some relief that he turned to another course of study under another adviser.

Still in his twenties, he shifted from dogs to plants, specifically the study of simple plant reactions called tropisms. These reactions were originally the work of the brilliant, irascible, and drug-addicted botanist Julius Sachs at the University of Würzburg. Sachs had enumerated a set of tropisms—defined as any directed response by an organism to a constant stimulus, for instance, the way an aspidistra or ivy plant will turn its leaves toward the window where the sun comes in— and, as Loeb's new mentor, guided him in their further study. Loeb learned from Sachs a practical spirit that was unlike that of other researchers, who wanted to find answers to theoretical and philosophical questions. Guided by that spirit, Loeb wanted to use tropisms to suggest that plants were diversely functioning chemical machines.

Loeb drew up a semiology of tropisms, collecting all the particular responses a plant, sessile animal, or insect makes to its external environment. These were his building blocks. Introducing tropisms one by one in his 1906 *Dynamics of Living Matter* (a summary of his work from the 1880s), Loeb first set forth heliotropism, or the attempt of a living organism—be it a single-celled blob or a very complex sea animal or plant—to orient itself in relation to light. Geotropism, chemotropism, galvanotropism, rheotropism, and stereotropism, the respective responses of organisms to stimuli of gravity, chemicals, electric current, moving retina images, and the "pull" or influence of solid bodies, rounded out Loeb's tropism toolkit. Throughout, Loeb emphasized their compulsory quality. The green plant *had no choice* but to move by the compulsive force of heliotropism, turning mechanically toward the light, aligning its leaves with the angle of the rays. The *Spirographis spallenzani*, "a marine worm which lives in a stony tube," oriented itself toward the sun in a manner akin to the plant's, except that in its case the tropism was channeled through the worm's immediate milieu. Each time the sun moved, the worm secreted an elastic layer on one side of the interior of its tube, causing it to contract toward the light source.[4]

Tropisms always began on the outside of the creature they affected, manifesting themselves through the involuntary workings of the response mechanism as a shifting, a twitching, a pulling, or a turning. Such machinelike creatures had no "inner" contents: no will, no striv-ings, no conscience of their own. Relentlessly, Loeb located any originating impetus outside the organism. He also refused to make anthropomorphic attributions: for example, he warned that, while observing a positively heliotropic insect (such as a moth) fly toward a flame, one may be tempted to believe the moth feels a humanlike emotion such as a fascination for light. But Loeb cautioned, "It seemed to me that we had no right to see in this tendency of animals . . . the expression of an emotion, but that this might be *a purely mechanical or compulsory effect of the light*, identical with the heliotropic curvature observed in plants."[5] Seeing plants as reactive chemical-machines allowed him to extrapolate directly to lower animals, even of the "free moving" variety. For Loeb, no preconceived idea of freedom—free will, free expression—should exist within the laboratory context. Tropisms were, at root, *machinelike* behavior, outside of the promptings of will, yearning, or desire. They had no secret unity with human feeling, and no delirious butterfly was drunkenly following the light.

IT IS STRIKING how ordinary tropisms are, in light of the extraordinary uses to which Loeb put them. They make up the humblest aspects of the daily life of an animal. Everyone knows these behavioral tropes—a plant swaying toward a window, a dog seeking a fire, a cat curling up in a basket. During his second apprenticeship, Loeb made these banalities into something dramatic, a kind of theater. He trained cockroaches through the clever use of simple tropisms: the insect's bilateral symmetry meant that a light shone on one side would cause it to move in the other direction. Equipped with this binary choice of movement either toward or away from light, Loeb could in effect control behavior. Troops of cockroaches marched in geometric array in Loeb's laboratory. In another case, browntail-moth caterpillars could be made to starve to death in a test tube, even when they were perched right next to their food, if heliotropism turned them in the other direction. "We can easily show that neither smell nor a special mystical 'instinct' leads the animals to the buds," he wrote, "as we are able to compel them by the aid of light to starve in close proximity to food."[6] In Loeb's dramas, elements of the everyday could suddenly verge on the grotesque or the amazing. This early tropism work soon led Loeb to experiment with heteromorphism, using the modes of geotropism, stereotropism, and heliotropism to rebuild an organism and transform its development and functioning. He created a two-headed worm (bioral tubularian), "any number" of which, he claimed, he could propagate—"if, for any reason, it were necessary," he added somewhat vaguely. The ability to make new forms also meant the ability to mass-produce them.

During the late 1880s Loeb's engineering standpoint became more explicit, especially in correspondence with the Viennese physicist and influential philosopher Ernst Mach.[7] From Mach he drew the strength to insist no true causes existed, no mechanical ideal, no "instinct," no "will," no "mystery," and above all no "metaphysics." By *metaphysics* he meant anything beyond what could be seen, described, or discovered. There were no busy bees or stalwart bugs. The purpose of this stripping-away was not to speculate on hypothetical mechanisms or inner states but rather to be able to predict and control behavior. To see was to cause; to see was to change. In 1890 Loeb wrote to Mach:

> The idea is now hovering before me that man himself can act as a creator even in living nature, forming it eventually according to his will. Man can at last succeed in a technology of living substance [*einer Technik der lebenden Wesen*]. Biologists label that the production of monstrosities; railroads, telegraphs, and the rest of the achievements of the technology of inanimate nature are accordingly monstrosities. In any case they are not produced by nature; man has never encountered them. But even here I go forward only slowly. I find it difficult not to lose courage.[8]

Man could be as a god, creating new forms of life out of living parts. This was a source of anxiety as well as of hope, for, as Loeb admitted, tinkering with creation was a dangerous business. (Consider Dr. Frankenstein's "filthy workshop of creation" and what issued from it.) Loeb, however, had a warrant to press on: he would be using animate rather than inanimate materials. His technology of living substance might create unknown beings—strange creatures never encountered before—but it would at least be anchored in nature.

AT THE START OF THE TWENTIETH CENTURY, the city of Chicago teemed with slums, pickpockets, foreigners, money, and enterprise, all of which influenced the type of science that was conducted there. When Max Weber visited around this time, he felt that the city was like a human body with the skin pulled off, entrails working for all to see. An early course catalog for the University of Chicago put a more dignified spin on it: Chicago was "one of the most complete social laboratories in the world." The work of its scientists made it feel like a laboratory within a laboratory. All was within the domain of experiment. The work of the multidisciplinary Chicago School of Pragmatism was unique in the world. However much its adherents differed, they shared an emphasis on recouplings, interactions, and progress: the environment acted and the creatures living within it acted back, in a constant interplay between things-as-they-are and things-as-they-are-becoming. Nothing was settled. The organism and its surroundings acted on and molded each other. The philosopher John Dewey, the psychologist George Herbert Mead, the biologist Herbert Spencer Jennings, and the zoologist Charles Whitman were advancing new ways of looking at such human and animal interactions.

In 1892, in nearby Iowa, a man named John Froelich had unveiled the first tractor. His farming machine, which had the power to reshape the environment, spurred the invention and use of many other technologies in agricultural production. Soon crops such as cotton and wheat were custom built to suit the machines that harvested them. Between 1900 and 1921 more than seven hundred R&D laboratories were created in the United States, along with many experimental stations for agriculture. Grain elevators and agricultural water towers rose to mark the landscape with new totemic structures. In these surroundings Chicagoans saw less a Hobbesian nature, brute and brutal, than a malleable one, tailor-made for what the historian Richard Hofstadter once called "the philosophy of possibility." Here was an environment where an engineering standpoint—toward crops, animals, buildings, or people—might go far.

Loeb eventually made his mark in Chicago. Having for years come up against the limits imposed by academic anti-Semitism in Germany, Loeb met and married Anne Leonard, a well-connected American, and began thinking of moving. At first he considered becoming a gentleman farmer in the fabled farmland of Indiana, where he imagined himself keeping a laboratory on the side. (Conversations with a visiting scholar from the Midwest had sold Loeb on the idea of its vast spaces and fertile fields.) With a baby on the way, however, he realized that his inheritance might not suffice to support a family, and so he attempted, for the second and last time, to train himself in a more mundane profession, this time as an ophthalmologist. Soon he again encountered boredom and despair, for he had "questions that I have carried in my head for years . . . if I cannot work on them I cannot live," as he told his new wife. They immigrated to the United States in 1891, he lacking command of English but she having a fortunate connection with Clark University. (Its president, G. Stanley Hall, was her father's cousin.) Loeb was offered a job at Bryn Mawr, despite the college administration's reservations over his Jewishness, which they believed might deter daughters of the best Protestant families from attending. Still, Loeb did well there, and within a year he was able to gain a position at the just-founded University of Chicago, where his arrival coincided, and in a few cases collided, with the new interdisciplinary paradigm just being advanced.

His great enthusiasm for experimenting made him an exemplar of the attitude Chicago loved, though the man and the city were not a perfect fit. Loeb chafed at American adherence to progressive evolutionism—the view that nature, although not in any particular form, was always working toward something and bore within itself a manifestly intelligent design. But still he came to feel that living there was worth it and gave him the chance to work in the Midwest's "primeval forest." Even as he was transplanted, Loeb's purpose remained clear: "to form new combinations from the elements of living nature," to "produc[e] new forms at will," to "produc[e] living matter artificially," to discover the "energetics of life phenomena."[9] In short, he sought the basic blocks for building life, nothing less than the "ultimate units of living substance"; but he insisted that these must be real components that one could actually work with. From 1895 to 1898 he studied physical chemistry in search of a language that would encompass all phenomena. Loeb then continued his work at the university during the academic year and at the Woods Hole Marine Biological Laboratory during the summer.

EGGS, FOR THE TURN-OF-THE-CENTURY SCIENTIST, were of great interest. They were like miniature factories for creating life, microcosms in which the stages of development unfolded. They held life's secrets in a neat package. Great debates raged between so-called epigenecists and preformationists: did the fertilized egg grow by responding only to cues from its environment, or did it also follow a "built-in" track determined by inherited instructions? Loeb opposed the preformationists, who saw the egg as carrying out a predetermined recipe, but he nonetheless felt there was some "germ plasm" inside the egg giving out orders. In this sense he predicted the existence of DNA many years before the discovery of the double helix.

To Loeb, now thirty-seven, the fulfillment of his goal of creating life seemed sometimes close at hand, sometimes distant. It was at this point, while carrying out experiments on marine animals' eggs, that he managed at last to fulfill one long-held dream. He invented artificial parthenogenesis, a technique that gave promise, soon, of an artificially produced mammal, followed by a human. The basic technique was an applied tropism—stereotropism, to be exact: a sea urchin egg was surrounded in its normal state by sea water, which provided a constant stimulus. When Loeb altered this environment by adding a mildly acidic solution, the egg began to divide and reproduce itself automatically, almost like a machine. It was "triggered" by the tropism. The discovery of this technique, at once relayed to the nation in breathless accounts, made Loeb a star. Novelists and newspapermen saw possibilities for test-tube-generated life, for women to have babies without men, for factory farming of domestic animals and children. Loeb, perhaps carried away, confided to a reporter, "I wanted to take life in my hands and play with it—to start it, stop it, vary it, study it under every condition."[10] To some he looked like a mad scientist, while to others—such as Sinclair Lewis, who featured in his 1925 novel *Arrowsmith* the noble yet irascible scientist Max Gottlieb, based directly on Loeb—he seemed a high priest of the laboratory whose presence inspired a kind of awe.

In 1903 Loeb was wooed away from Chicago by the University of California, whose regents wanted to make an international name for themselves by recruiting the nation's most famous scientist. They agreed to give him everything he wanted, including no teaching or administrative duties, a decent salary, several junior positions to fill as he liked, and a special laboratory in New Monterey, near Pebble Beach, where he could live and conduct experiments to his heart's content.

Meanwhile the University of California was also promoting Luther Burbank's strange experiments. Just as Loeb used his laboratory to craft new life-forms, Burbank drew crowds that clamored to see his fruit, flower, and arborial creations at his Experiment Grounds. The two scientists also shared a distaste for the too-well-trained academic. Burbank once complained to the president of the Carnegie Institution about a man sent to oversee his work: "It seems to be almost necessary to perform a surgical operation before some fixed impression can be removed to make way for another, but once convinced by an overwhelming number of facts he at once admits that that is the way he always thought it was."[11] Both cultivated an attitude of receptivity, finding that fixed impressions got in the way.

Although Loeb's later efforts were not as sensational as his earlier work on virgin births, he continued to pursue proof of the physicochemical basis of life. His conviction that a firm understanding of the life processes of the simplest creatures was the basis for moving on to humans took root among a widening circle of scientists and social thinkers, such as the heterodox economist Thorstein Veblen; Gregory Pincus, the inventor of the birth control pill; B. F. Skinner, then a graduate student doing his dissertation on tropism in ants; and Loeb's literary admirer, Theodore Dreiser. Further research was called for, and the end goal remained controlling human behavior.

LOEB'S PROJECT and its aftereffects (the people he influenced, the programs he inspired, and the cult-figure status he attained) raise questions about the ethics of treating living things as machines. Some critics felt that Loeb, in pursuit of his goals, had ignored everything important and interesting about life, its particularity, unpredictability, messiness, and passion; perhaps its divinity; certainly its soul. And his work did seem to pave the way for the recombination of more than sea urchins and starfish—humans were next in line. In this sense his experiments have often been labeled "materialist" or "mechanistic" or "reductionist" or simply "bad." They have been connected with the intellectual views that separate mind from matter (often traced to Descartes) and the scientist from the natural world (often traced to Bacon, who wrote of learning to "torture nature for her secrets"); and with the relentless buying and selling of the life processes, the death processes, and everything in between (traced to the economic ravages of capitalism). All in all Loeb's work appeared to be a damning slide toward the domination and enslavement of all of nature.

As William James once said, certain scientists (he called them materialists) sought to "defin[e] the world so as to leave man's soul upon it as a sort of outside passenger or alien."[12] Loeb appeared, to some, to be one of them. In denying that humanlike qualities of will, yearning, and desire were operating within tropistic organisms, Loeb reserved "will" for the human actor, the scientist. Those who followed in his footsteps, inspired by his rigor, his progress toward creating life, and his seemingly mechanistic viewpoint, eventually went on to deny a "will" to human subjects in the laboratory as well.

Loeb's science paved the way for others to see that life itself is subject to design. It was at Chicago that the later crystallizer of behaviorism as a "movement," John B. Watson, as a graduate student, came under the special influence of his biology and physiology teacher, Loeb. Watson, watching Loeb's work with interest, took his insights to the extreme and eventually mapped out a science of behavior that could conceivably explain everything—from a houseplant facing the sun to a philosopher writing a tome—in terms of stimulus-and-response reactions. So Loeb's turn-of-the-century experiments on tropisms sketched out a vast matrix of stimulus-response mechanisms that later brought the engineering of human fears and desires into the realm of possibility. Sometimes the behaviorists put it rather crassly, as when Watson boasted of being able to take a baby and "build" any type of man (by evoking and recombining the infant's conditioned responses to fear, love, and anger). But the basis of human engineering, at least as Loeb and his fellow pragmatists liked to say, was in a quality of observation, a style of looking, and thus a style of inquiry. The world could be altered by the way one looked at it, and in the confined space of the laboratory, the new techniques of looking at objects, people, and phenomena of nature could be tried out intensively.

Through his laboratory practices, Loeb could build new life-forms out of functional parts of sea worms, houseplants, and hydras, and this betokened other changes as well. To design new holding places, mazes, and conduits for herding, molding, and shaping humanity (through tropisms, conditioned responses, and other techniques) was indeed to take life into one's hands.

CHAPTER 2

Running the Maze

IF YOU WENT to Grand Central Terminal in 1928 and ducked into the adjoining Graybar Building, you would have arrived in the lobby of the world's largest office building. If you had then taken the elevator up to the offices of its largest tenant, proceeded through the art deco doors marked "J. Walter Thompson Advertising Agency," and continued down the hall past the vast executive dining room with its fifteen-foot-high fireplaces, you would have found—opening yet another wrought-iron door—the singular person of John B. Watson, vice president. Here sat the great icon of the science of behaviorism, an émigré from the state-of-the-art laboratories of Johns Hopkins University to the high-powered selling rooms of Madison Avenue. Visitors often observed him at his desk, ensconced in lush surroundings designed by the well-known futurist Norman Bel Geddes. People were eager to meet this legendary scientist who had plumbed the depths of the human psyche and found them to be less interesting than previously supposed. Watson, with characteristic brisk authority, had argued that if scientists could only strip away their assumptions about the existence of an inner soul or an essence of mankind, they could predict exactly how human beings would function in different situations.

This was the project to which Watson had devoted his academic career, keeping the public abreast of his discoveries through an unflagging series of magazine articles in *Harper's* and *The Saturday Evening Post*. Now, amid the opulence of his office, he was not only an established talent in the field of advertising but also a symbol of the powers of science brought to bear on the exigencies of business. He had the aura of a guru and knew more than anyone else about the finer points of what happened to people when they were poised at what we would call today the "point of purchase." Watson claimed there was nothing he could not sell. He made good the claim not only in his career as an ad man—strategizing campaigns for "inhalable" cigarettes and underarm deodorant just as these new products were first entering the mass market—but also in his capacity as a man of science.

Watson's New York position might have marked the apogee of his career. His influence was unparalleled, his name widely known, and he was better compensated than any professor. A robust and handsome man, he lived in splendor on a Connecticut horse farm with his wife, a former lab assistant, Rosalie Raynor, and their two young sons. But in fact Watson's niche at the J. Walter Thompson Agency was something of a comedown. Not ten years before, he had known a different and, for a man of his inclinations, more gratifying kind of glory. He had been hailed by one of the greatest philosophers of modern times. In 1919 Bertrand Russell, unequivocally a man of genius, had written him an admiring letter and made Watson's work the focus of an entire book, *Philosophy*. The gist of Russell's argument was that one could no longer consider traditional philosophical problems without taking into account the science of behaviorism. And Watson's work on the subject impressed him.

The behaviorism Watson set forth was simple but revolutionary: instead of trying to study what could not be seen (such as mental states, hazy emotions, or the ever-elusive human soul), the psychologist was to focus solely on what could be seen and measured—that is, on behavior. The resulting science was on the verge of radically remaking the age-old question of the relationship between mind and matter, which, at least since Descartes, had become the question of whether thought or the objects of thought were the primary reality. (Idealists believed that thinking or consciousness was primary; materialists fastened on matter or "things"; and philosophical and psychological peregrinations had been stacking up on the topic for centuries.) Watson's answer was a brilliant and breathtaking shortcut—*focus on what we can know scientifically and ignore the rest.* Other great thinkers were also following Watson's every move. His former teacher John Dewey, perhaps the greatest American philosopher of the day, counted himself a "well-wisher" of Watson's behaviorism, while the cutting-edge psychologist George Herbert Mead impatiently awaited the results of each of Watson's experiments. (This was heady stuff for a man who had not taken to philosophy in school and who had once said of Dewey, "I never knew what he was talking about."[1]) Graduate students flocked to Watson. In 1919 he was fresh from a year's tenure as president of the American Psychological Association and was editor of the most prominent journal in the profession. Some called him a second Moses (parting the waters and leading his young psychologist followers from the hopeless land of philosophy to the promised land of a true science of humankind). Having begun his experiments with animals, he was busy transferring the results to humans. His horizon seemed unlimited.

But in 1920 a scandal overtook him, and he found his career abruptly interrupted. He shared with the great pragmatist philosopher Charles Sanders Peirce the fate of summary dismissal from Johns Hopkins, and accordingly from academic life altogether, for the crime of sexual dalliance. (Both men met and later married women who were not their wives while teaching there, and in both cases the overlap between first wife and girlfriend was judged unseemly.) However, unlike Peirce, Watson landed not in penury but on a firm footing of prosperity. Still, the academic world no longer attended to him. The science of animal and human behavior went on without him, indeed so successfully that the inheritors of his technique and bearers of his science (for it was Watson who coined the very term *behaviorism*) effectively wiped the slate clean of him. Usually he was remembered as closer to a shill than as a significant scientist, and although he may have been given the obligatory mention in journal articles, his work was modified by the adjective "crude." For academics, who quickly moved to subtler ways of developing the principles Watson had brashly set forth (and who were not keen to acknowledge a Madison Avenue forebear, sexual misadventures or no), he was like an embarrassing relative one would rather not invite out.

Watson's early scientific work on the behavior of animals under various conditions formed the basis not only for behaviorism proper but, later, for neobehaviorism, modified behaviorism, operant conditioning, learning theory, and a slew of behavior-modification techniques—to say nothing of much of clinical psychological work today, the self-help movement, and every advertising technique. Classical behaviorism may have gone out of style, but its procedures, especially the running of albino rats through laboratory mazes, proliferated. It set into motion practices and assumptions whose applications have had no limits. We live today among its many fruits, without realizing, much less acknowledging it. For example, the coauthor of the famous *Bell Curve*, Richard Herrnstein, who trained as a behaviorist at Harvard, could not have made his controversial argument about how race in human beings is linked to IQ without doing extensive work on pigeon behavior in the laboratory. More broadly, the ubiquitous use of behavioral conditioning within shopping environments, office parks, managed-care outlets, and multiplex coffee shops is also a direct fruit of behaviorism.

Throughout his career in and out of the laboratory, Watson cemented a specific equation: the activities of animals under experimental conditions were equivalent to human activities under all conditions. Understanding animal behavior was the key to understanding human behavior in all its forms, even the "highest." This equation allowed the social sciences—and as a consequence social life—to move in a certain direction. It validated the use of animals as experimental stand-ins. Watson believed that the twitchings and turnings of creatures made to run countless trials through cleverly built laboratory de-

vices would provide scientific insight into the intricacies of human behavior and the most deeply hidden parts of the human psyche. How
exactly did the rat-in-the-maze (to use a convenient shorthand for
Watson's innovations) come to dominate psychology and eventually
the mainstream social sciences, philosophy, and the vast areas of public persuasion? That dominance lasted for decades and eventually suc-
cumbed only to its own success—which is another way of saying it did
not succumb at all.

THE WHITE RAT was the first species in history to be domesticated
for science. In the 1860s French biologists began keeping them in laboratories to study breeding patterns. At that time no rat of any type
had been domesticated for very long. In the early 1800s *Rattus rattus*
(black rats) and *Rattus norwegians* (brown) were being kept and bred
as rat baiting took off as a sport. Whether it was cultivated for fighting
or for study, the rat had certain advantages. As an early rat studier observed, lab rats are "small, cheap, easily fed and cared for; and best of
all, when placed in revolving cages, they spend most of their time . . . in
running."[2] They were warm-blooded but did not inspire affection. Nobody much cared what happened to them (you could, as Watson once
remarked, castrate, shock, poison, blind, drug, or "surgically interfere"
with a rat, and be perfectly within your rights). The first rats to come
to America as objects of scientific study, according to record, were
imported in 1892 by the émigré Swiss psychologist Adolf Meyer for
his laboratory at the University of Chicago. From there, rat colonies
spread.

How rapidly they spread had something to do with the growing
hopes and dreams for human engineering and with a fervor to conduct
laboratory experiments to advance them. The laboratory setting con-
ferred scientific authority that made all kinds of manipulations acceptable and that allowed Watson's equation to be acted upon. (At the same
time it enforced the separation between animal and human, for clearly
what could be done to a rat was not at all what could be done to a person.) Even though the public, as the twentieth century got going, was

more and more inclined to decry research conducted on rats as cruel—at least when the experiments and their results made the newspapers, which was not very often—scientists found it beguiling. The pain felt by animals was a component of essential research, a necessary by-product. According to the historian Philip Pauly, "biologists were con-vinced that what seemed brutal to middlebrow women was in fact a sign of moral refinement."[3] Psychologists as well as biologists main-tained that their experiments with animals were morally refined. On the one hand, they were exigent, for they were central to the important task of understanding the totality of human function. On the other hand, they were elegant, for they made animal pain a substitute for human pain. The lay public failed to understand this argument, and experimentalists, finding it better not to speak of certain activities, increasingly confined themselves to enclosed spaces where their stan-dards could not easily be challenged. Laboratories became spaces of retreat.

Rats were introduced into the first miniature mazes at Clark Univer-sity in the late 1890s, a decade or so before Freud gave his debut Amer-ican lecture there. Two researchers, Linus Kline and Willard Small, wanted to create a more natural situation than Edward L. Thorndike's renowned "puzzle boxes" from the same years. The puzzle box was a device requiring a solution, a wire-mesh contraption with various pul-leys and levers that the creature inside had to operate to exit and get some food. The Clark researchers wanted something less devicelike, more like the burrows of wild rats. Small then reconstructed an envi-ronment in which rats could live, with access to food at the center. He designed it after the Hampton Court Palace maze, a trapezoidal hedge maze that dates in its present form to 1690 but that was probably built in the early sixteenth century by the first owner of the palace, Cardinal Thomas Wolsey. It is not clear why the two researchers chose this par-ticular design, for there were many such mazes all over Europe, relics of an aristocratic vogue for running among hedges in pursuit of amusement. In these early experiments, the Hampton Court maze-in-miniature served as a pied-à-terre in which rats lived and took exercise when they were not being studied. Only later, under Watson, would the maze become itself a problem that had to be solved. The Clark program eventually died off, and by the turn of the century Chicago was the only place left with rats in labs. There the tradition was maintained unbroken until the great surge of rat research in the mid-1910s and early 1920s.

This surge was the almost-single-handed result of the work of Watson, whose dissertation research, beginning in obscurity in the fall of 1901, was the first of its kind. Although he acknowledged his debt to Small, Watson contrasted his own rats with the other's "timid and flighty" ones: "The rats used in the experiments reported in the present paper were raised by the writer in the laboratory and were exceedingly tame, as shown by the fact that they were not disturbed by handling and at once investigated all new objects in their neighborhood."[4] He built his own mazes, which were much simpler than the Clark researchers' Hampton Court design, and set a standard in the field, for they tested the rat within the maze itself. Watson's other ingenious devices had rats burrowing through sawdust to get to an opening, pressing a lever to open a door, or walking out to the end of a gangplank-like inclined plane in order to trigger a spring-latch. Each one led to cheese or bread, the usual bait, and all the while Watson carefully observed and measured the rats' ability to learn.

In 1906 less fortunate rats had one or more of their sense organs eliminated. How would a rat navigate a modified Hampton Court maze with its eyes put out or its middle ears blocked, its olfactory bulbs extracted or its whiskers plucked, or without any sensation on its paws? Could the creature proceed? Would it? The answer was yes: using feelings in their muscles and their guts—which Watson called "kinesthetic sensations"—the rats could still move around the maze with some success. Aside from its scientific merits, about which there remains some controversy,[5] the experiment was affecting and played out like a story. Wandering in a strange and obscure landscape that had once been familiar, a blind, deaf, or muffled rat managed to make its way through the dark unknown to safety. The story was too affecting, perhaps, for despite the reassuring denouement there was a public outcry about the cruelty inflicted on the rats, the thrust of which was summed up in a *New York Times* editorial, "Torture to No Purpose."[6] Unschooled reactions from antivivisectionists, middle-class ladies, and sensation-mongering journalists annoyed experimenters, who insisted on the scientific merits of their work. From the time of these early experiments Watson felt that the basic trial-and-error mechanisms revealed by the rat work could just as well apply to humans, but as he told it, his was just a "voice crying in the wilderness" and no one heeded him for a good five years. It took another five years for the greatest psychologists and philosophers of the day, along with the public, to take note. Watson's much-maligned and often misunderstood role in the advance of this experimental tradition helps explain how this upstart subsubfield of psychology, with its unusual laboratory method of equating animal to human behavior, eventually became central to the assumptions and architecture of modern life.

JOHN BROADUS WATSON was a classic up-by-the-bootstraps American type, his life a testament to self-propelled upward mobility. He came from a part of South Carolina so obscure that the nearest school, which required a lengthy commute on foot from the age of six, was two miles away in a place called Reedy River. His mother, Emma, an evangelical Christian, named her first son, the fourth of her six children, after the Southern Baptist preacher John Broadus with hopes he would follow his example. At age twelve, Watson moved with his family to Greenville, a much larger town. Their peregrinations sometimes did and sometimes did not include their father, Pickens, a wanderer, brawler, and ne'er-do-well from well-to-do stock. The eldest son, self-described as "antisocial," had few close friends growing up and took after his father. Watson's autobiography is to the point: "I used to have a friend by the name of Joe Leech with whom I boxed every time my teacher left the room, boxed until one or the other drew blood."[7] Throughout this document, he is frank about his strivings, jealousies, pettiness, and willingness to do hard work, all written in a tone of plain-talking one-upmanship.

Despite his extreme poverty and his pugilism, Watson made it to

Furman University, a Baptist school, where he hoped to study philosophy and psychology. Right away the experience turned him off to philosophy (which "wouldn't take hold") and college life itself (its "failure ... to mean anything to me" and more generally its encouragement of "a prolonged infancy"). He was impatient with the anxieties and self-absorbed worries of the middle-class students who surrounded him. Perhaps his disdain for introspective mollycoddling had something to do with his lifelong scientific program to deny the very existence of introspection. (Thought, he contended, did not exist except as non-verbalized speech.) At any rate, some years later, on the verge of finishing graduate school and while holding down several jobs, he had a nervous breakdown that lasted for several weeks, caused him to awaken at 3:00 a.m. to walk for miles, and gave him, he reported, an unsought "understanding of Freud."

Despite these perhaps inauspicious beginnings, Watson went on to do innovative work as a student and young faculty member at the University of Chicago. He studied biology and physiology with Jacques Loeb, from whom he learned, he said in an understatement, "that all research need not be uninteresting." Once again, however, philosophy failed to hold Watson's attention, even the teachings of John Dewey, whose serene personal presence and vast intellectual gifts normally drew students to him. Watson preferred the brasher, unconventional anthropologist W. I. Thomas, who befriended him, and the psychologist G. H. Mead, whose seminars he took. Rounding out his mentors was James Angell, a founder of the New Psychology, who studied the way organs function rather than their structure. Angell, who later became the president of Yale University, was impressed by Watson's ambition and gave him a job as an assistant janitor cleaning apparatuses in the laboratory. Watson also delivered newspapers, waited tables, and cared for Professor H. H. Donaldson's white rats—and this was where a historical concatenation resulted.

The rat-in-lab tradition, which had been kept alive at Chicago but was otherwise foundering, was revived again by this young and penurious graduate student who, getting to know the rats and having a country boy's way with animals, started thinking up new and ingenious experiments to do with them. Watson felt he was blazing new ground: "On the maze work I felt a certain independence," he recalled in his autobiography. In 1903 he became the youngest person to receive a Ph.D. from the University of Chicago, and although he castigated himself for having been preceded in securing a summa cum laude degree by a female student two years before, he was offered the chair in psychology at Johns Hopkins at age twenty-nine.

DURING THE NEXT SEVERAL YEARS, Watson labored long days and nights at Johns Hopkins, refining his experimental method of close observation and repeated testing of hypotheses. He spent several months on the Dry Tortuga islands off the coast of Florida with the native terns, then returned to the laboratory to study sundry wild and tame creatures as well as, of course, albino rats. In 1908 working with animals was generally a guaranteed ticket to obscurity within the field of psychology, which was, after all, supposed to be about the mind of man, not beast. Such work usually attracted only eccentrics and dalliers interested in the curious abilities of certain talented cats or foot-stomping horses. But within a decade animal experimentation would become the cutting edge and most creative part of an American science that was itself growing rapidly. Behaviorism became the newest of new psychologies.

During this time, as Virginia Woolf pointed out, the world as people had known it abruptly ceased to exist. "On or about December 1910," she wrote, "human character changed." People, places, and things all seemed to be remaking themselves along new lines. The sense of living in a streamlined manner, not just a life but a style of life, had taken hold. A radical spirit—anything can be made or remade!—applied not only to things and buildings but to the people who made them or worked in them. Great capitalists such as John D. Rockefeller and Andrew Carnegie set up foundations devoted to social engineering in which "the possibilities of social experimentation are to be kept constantly in mind."[8] John Dewey's Laboratory School in Chicago remade early education as an experiment taking place under experimental conditions. (Dewey believed that education was not a matter of rote learning and endless drilling but should encourage learning by doing.) Designers built ergonomic gadgets to occupy futuristic spaces, and architects experimented with the "lived environment," modeling apartment buildings on the clean functionality of modern factories in Buenos Aires and Buffalo. An American movement called Technocracy, Inc., inspired by the economist Thorstein Veblen's views on efficiency, declared without irony a "new era in the life of man."[9] Other esoteric, forward-looking groups begot themselves, with names such as the New Machine, the Technical Alliance, and the Utopian Society of America, and devoted their energies to molding the future, while European movements such as Constructivism, Purism, and the middle Bauhaus were equally enthusiastic, if more arty, about the cult of the machine. Some dreamed of engineering themselves; others dreamed of engineering the rest of humanity.

But these engineered selves and liberated spirits, so firmly announced and eagerly sought after (at least by those forward-looking types inclined to seek such things), were also a cause for concern. The impetus "to remake the world to human specifications," as the historian Dorothy Ross has characterized it, could also mean remaking human beings to the scientist's specifications.[10] What were the most extreme implications of these projects? A hooking up of human and machine? A fleet of automaton workers enslaved to the dark satanic mills of industrial production, as portrayed in Fritz Lang's 1926 film *Metropolis*? Where would hopes for reason-led-by-science be fulfilled?

Dark inklings of future difficulties aside, experiments in human engineering continued in the big cities and universities of Europe and America. Especially in science there was a strong movement to investigate human beings as life-forms within an environment. The goal was the reconfiguration of human behavior and eventually all human capabilities, so that man-within-the-environment was no longer a fact simply to be accepted but an assemblage to be changed. For this scientists needed two things: laboratory space to do the experiments, which was no problem to obtain, and laboratory subjects, which most certainly were—particularly in cases in which no direct therapeutic benefit for the participant was likely to accrue. Human engineers and other experimenters had to rely on the substitution of animals for human beings.

Consider a key experiment by Watson. In 1915 he heard rumors of Ivan Pavlov and Vladimir Bechterev's groundbreaking work in Russia on the conditioned reflex in dogs. Their first innovation was to insert a fistula or "window" into the body of a dog and observe its digestive system at work while the animal was alive (resulting in a Nobel Prize for Pavlov in 1904). The second innovation was to attach a test tube to the dog's salivary glands, as a way to measure the dog's innate response to the arrival of food. They then found that an "unconditioned" stimulus (meat) could be used to condition other, unrelated stimuli (a bell, a light, the scuffling of his master's feet). Initially the dog drooled right before he ate, but later he also drooled when a red light flashed and a tiny carousel played music. Later still he stopped drooling at either light or music through what was called "experimental extinction." Just about anything could be conditioned to just about anything else. Although the Russians' work was not yet available in English translation, Watson managed to secure a rough account, and he immediately saw the revolutionary implications of being able to recondition the simple behaviors of an experimental animal. Watson set out with alacrity to broaden the focus of such research. In 1916 eleven human subjects (including one child), one dog, seven chickens, and a great horned owl took part in an experiment. The eight-year-old, Watson noted, cried when shocked with an electric current and would continue only with the promise of "a moving picture show" afterward; adult subjects, when burned with cigarettes, also established aversive reactions rather quickly. At this point Watson harbored hopes for continuing laboratory experimentation on the sick, the young, and wild or large animals. A photograph of an experimental owl ensconced in elaborate machinery bore a caption that is reassuring in its matter-of-factness, yet strange as well for that very reason:

Method of obtaining respiratory reflex in all birds. The great horned owl is shown resting comfortably in a padded wooden saddle. Underneath the floor of this apparatus Rouse's respiratory apparatus is shown, sliding on vertical rods. A V-shaped button is shellacked to the receiving tambour, which is adjusted lightly against the bird's chest. The owl's feet are attached to a punishment grill.[11]

An almost surreal expectation of scientific reach and grasp asserted itself, as if all of nature, not just its domesticated spheres, could be swept up into laboratory research. In these same years another pioneer in animal research, Robert Mearns Yerkes, also worked to extend the range of possible subjects; although usually known as a primatologist, he conducted trial-and-error studies on reptiles, amphibians, crabs, squirrels, and "dancing mice" (hereditary oddities that displayed constant restless movements). These researches had a Noah-like quality, except in reverse: the experiments were not to save the creatures brought on board but to save those left outside. Watson believed that the mass of humankind, foundering in rising waters, would be the true beneficiaries of a laboratory science for engineering behavior.

Within a few years this willingness to extend research to farther-flung species died down, and researchers settled on a few particular ones as acceptable for experimentation. The field of behaviorism was on its way to consolidation; its procedures became standardized, its assumptions widely shared. In labs and at research stations scientists could do to animals what they could not do to humans. After 1919 one rarely saw owls or eight-year-olds in behavioral experiments. Certain species of animals took a place on the laboratory stage for psychological, social, and psychical research.

In the 1920s and 1930s an unofficial rule was established: albino rats, certain apes, and the guinea pig would be the paradigmatic, archetypal lab animals. In particular, specially bred lab rats were de rigueur, the obvious choice. Experimentalists no longer needed to justify the use of rats over other creatures, for scientific consensus now had it that they were somehow the most standardized and standardizable of subjects. Coming across the occasional written justification of the lab-rat-as-stand-in, one would hardly guess that its use was only as old as the new century was young. Scientists had a sweeping confidence in using the animals, as B. F. Skinner later explained: "In the broadest sense a science of behavior should be concerned with all kinds of organisms, but it is reasonable to limit oneself, at least in the beginning, to a single representative example. Through a certain anthropocentricity of interests we are likely to choose an organism as similar to man as is consistent with experimental convenience and control."[12] Convenience and control: Skinner perfectly summed up the necessary qualities, having the estimable ability to put elegantly and economically what other researchers left unspoken but were feeling and acting upon. They chose rats because they considered them close enough to humans to be representative but not close enough to be disturbing.

The choice of "problem situation" became standardized as well. Ten years after Watson's groundbreaking dissertation, the maze—which throughout Western history had served as a potent literary and religious symbol of the difficulties of finding meaning in what Kant called "the labyrinth of evil into which our species has wandered"—was teeming with a new kind of activity. An army of clunky prototype mechanisms (such as Thorndike's problem boxes, Richardson's Jumping Device, and Dr. Yoakum's Temperature Apparatus) dropped out of research, and the maze triumphed in its various permutations: T-mazes, modified Hampton Courts, labyrinthine confections with variously designed corridors and alleyways, traps, and false passageways. Mazes won out because in a sense they were the most general, the most representative, and the most perfect models available of the original problem situation, life itself. They were a shorthand way of asking, "Why does the self behave as it does?"[13] An animal within a maze was faced with choices, confusions, blind alleys, and difficulties. Finding himself hungry (that is, "motivated") and in a place where dangers lay all around (electric shock, cold, poison, thirst), the rat encountered the twists and turns of what Milton in *Paradise Lost* called "mazy error" and what behaviorists called "choice-points." The maze had long stood for the struggle to find one's way when the truth was elusive and the way fraught with monsters and despair: Theseus had killed the Minotaur in the labyrinth; the Christian Wayfarer of *Pilgrim's Progress* sought God in one; and Nietzsche begged to be lost in one.

In the laboratory maze this layered knowledge was both evoked and not evoked, like an emperor fully clothed who was supposed to be edifyingly naked. A rat might reach the goal or not, but the quest was no longer real. The point was not to succeed or to fail but merely to demonstrate the operations of a mechanism (turning aside here, avoiding a shock there) so that the scientist could observe the ever-unfolding web of stimulus-response. Since the "subject" running the maze was a stand-in, a throwaway, the scientist, having stepped aside to look on from above, was now in God's position: instead of a Seeker in search of meaning or absolution, one had a Scientist in search of mechanism. That rats became frustrated in mazes or gratified when rewarded might seem clear, yet the science of behaviorism was made to factor out such things as emotional states, innerness, subjectivity, and the unconscious and reduce activity to a series of blind mechanisms. For in the maze behavioral mechanisms could be found, sketched out, delimited, and experimented with; one could then establish with unimpeachable certainty that the same mechanisms also operated in human behavior.

Watson's equation was soon so successful that its truth became self-evident: if you could do it with rats, you should be able to do it with humans. Soon vast programs of energetic experimenters devoted themselves to using rats in mazes. The maze promised a great deal, for it provided the design for the new "human maze." If an array of basic reactions could be found and isolated in laboratory animals within the laboratory maze (as Loeb had done, as Watson was doing, and as Skinner and others were about to do), then it stood to reason that the same reactions could be found among ordinary human ruminants in the world at large. Scientists could train a lab rat to take a particular path to a desirable goal, and likewise (potentially) a human being.

Above all, behavioral scientists turned to rats in mazes because the new life-forms they hoped to engineer had to have an organic or living basis. They would not be mechanical through and through, but would be made of building blocks of living functions. This project reversed the premise of the mechanistic automata that had so fascinated some medieval Europeans, in which all-machine parts eerily "came to life" in performing a miraculous task. For example, the thirteenth-century French architect Villard de Honnecourt built a mechanical eagle that always faced toward a person reading the gospel; Robert of Artois devised *"engiens d'esbattement"* or machines for fun, among them a group of mechanical monkeys with horns attached, as well as an elephant, a goat, a hydraulic stag, and a carved tree with birds spouting water. The twentieth-century idea was to make out of living parts a machinelike creature filled with mechanical rhythms that were also somehow natural.

DUE TO THE SUCCESSES of Watson and others, animal experiments were carried out in American psychological laboratories with a fervor that is at first hard to understand. In the decade between 1919 and 1929, the number of rat-in-maze studies almost tripled. In their eagerness, experimentalists carved impromptu laboratory spaces out of administrative offices, basements, even closets. And then, just as the labs were being built, the albino rats reared, and the mazes designed, Watson produced a momentum-generating document: his manifesto. In the early spring of 1913, Watson, a brash thirty-five, gave a speech declaring behaviorism's arrival. The unveiling coincided with the 69th Regiment Armory Show, often seen as the start of full-fledged modernism in American art, where works by Vassily Kandinsky and other abstract European artists shocked the gathered throng of New Yorkers. Speaking at a venue uptown, Watson also struck a modern note and gathered a surprisingly large crowd of his own. His speech, "Psychology as the Behaviorist Views It," got right to the point. In the first two sentences he pared away the previous five hundred years of knowledge about the way to study the human mind and offered behaviorism as an alternative, declaring it to be a "purely objective" and inherently experimental branch of the natural sciences. As such, it had a single purpose: "the prediction and control of behavior." In pronouncing his approach a real science—perhaps the only one capable of dealing with human data—he asserted as a necessary corollary that its practice would lead to the engineering of all that humans do. The third sentence did away with consciousness—or rather said that it played no part in the data. And the fourth, getting even more to the point, recognized "no dividing line between man and brute." The rush to the lab was now explained: it was not just the stumblings of hungry rats in mazes but the future of the human species that was at stake.[14]

Watson owed much to his forerunner Jacques Loeb, but he threw aside his teacher's strong belief in the importance of ever-keener observation, in seeing things as they are. Watson was more interested in seeing things as they might one day become. Not that Watson was an unskilled observer, but as his career took off, he was more than willing to accommodate his data to his program rather than the other way around. He was capable of out-and-out distortion, of misrepresenting his results through amplified claims. Stripped of subtlety and scientific caution, Watson's behaviorism took hold and, in the various modified forms it subsequently assumed, would not yield for another fifty years or more.[15] Watson called out for a new approach in his manifesto:

> I believe we can write a psychology, and . . . never use the terms conscious-
> ness, mental states, mind, content, introspectively verifiable, imagery, and
> the like. . . . It can be done in terms of stimulus and response, in terms of
> habit formation, habit integrations and the like. . . . My final reason for this
> is to learn general and particular methods by which I may control behavior.

Others in the field of early behaviorism, such as Karl Lashley, Herbert Spencer Jennings, and the Johns Hopkinsites in his department, were interested more in understanding than in control, but Watson carried the day. Many, especially the young, the innovative, and the ambitious, agreed with Watson. Some said he had shown the way, at last, to a unified theory of mind and body.

STILL, THE POINT OF BEHAVIORISM was not to lose oneself in mazes of theory (not even in the most ambitious of unified theories); it was to *act* in the world. Behaviorism would be the avenue for bringing about true social and behavioral change. In the fall of 1916, inspired by the thirteenth-century experiments of the Holy Roman Emperor Frederick II, Watson started experimenting with babies. According to legend, on the instructions of the emperor, several babies were nursed on an island, where they were exposed to no language and no culture, to see how they would turn out. The babies not only failed to speak Greek or Hebrew, as had been hoped, but they all died. Social scientists typically rued the impossibility of conducting any such human experiments themselves, but Watson went ahead. He originally turned to babies because their actively developing senses would allow him to study simple responses to simple stimuli as these reactions unfolded. At the Phipps Clinic in Baltimore, he experimented with several babies ranging in age from three months to a year, including a nine-month-old named Albert who, as the result of tests run in the winter of 1919–20, would become an enduring part of psychology's history. From these neonates, who either had no mothers or whose mothers were working in the hospital, Watson compiled an index of reflexes that were present at birth or soon after, wondering whether they could be built upon. How many trials would it take to stop an infant from reaching for a candle flame? Was a six-month-old naturally afraid of living furry animals? (The tests were done in a breathtakingly straightforward manner, considering the tender ages of his research subjects: Present a burning candle to a curious infant and see what happens, and how many times it happens. Introduce different animals and ascertain whether the infant fears them naturally or only after repeated traumatic experiences. Make a hissing noise; observe results.) Watson discovered that newborn babies showed no fear of the dark, although they did have a basic inborn fear of falling and of very loud, clanging noises. On the other hand, they could be made to fear the dark, and many less likely things, as the case of Little Albert was soon to prove.

In his early tests, Watson attempted to simulate a thunderstorm in the laboratory by using a sudden flash of light from a heliostat together with a loud sound. Despite unprepossessing initial results and the dawning realization that babies were more difficult to work with than he had supposed, he moved in a new direction: toward examining the emotions. He began to home in on the conditioned reflex, seeing it as a hub where emotions get attached and detached. He believed that the conditioned reflex was the root of all emotional life in all human situations, acting upon the three basic emotions of fear, love, and anger—although fear seemed always to interest Watson the most.

This was the backdrop for the Little Albert experiment, which confirmed and dramatized the emotional power of conditioning. Watson exposed the baby to a loud noise (made by clanging a steel rod with a claw hammer just behind his head) every time he touched a white rabbit, an animal that had at first delighted him. Soon the baby reacted with fear to the rabbit even when the loud noise was not administered. Eventually other white, furry objects would elicit this response—at least, this was the claim Watson made, a claim that entered folklore and generations of textbook accounts of "stimulus generalization." In fact, the infant did not regularly demonstrate fear in response to any particular object, and the experiment was successful mostly in showing that Albert, when badgered sufficiently and not allowed to suck his thumb ("again and again . . . we had to remove the thumb from his mouth before the conditioned response could be obtained"), finally placed his hands over his eyes and whimpered in response to a Santa Claus mask, a sealskin coat, a dog, and Watson himself. The point, for Watson, was that fear-based conditioning such as Little Albert's, rather than love-based, was the most common and powerful force shaping a person's social life. Albert's mother, a nurse at the hospital, withdrew her son rather abruptly from the experiment at this point, before further trials could be conducted. Watson never attempted to decondition the baby, and it is not clear what happened afterward to him.[16]

Actually, with its questionable results, use of only one subject, and manifest procedural weaknesses, the experiment dramatized more than it confirmed and served mostly to supply future generations of students with photographs depicting, purportedly, how a response can spread from one stimulus (a white rabbit) to another (a fur coat). And yet Watson's renown was only enhanced by the Albert experiment. His fame secure, Watson knew no bounds for his ambitions. The conditioned reflex was the mechanism for which he had been looking, and a bridge to the emotional life. Now control was that much closer. He began to conduct experiments on subjects' learning and performance under the influence of hypnosis, alcohol, and drugs. Always in his mind was the beckoning vision of control over a person's development. He dreamed more ambitiously of "an experimental farm for babies" of different racial groups. Long after he had left academia the dream persisted, taking on an elegiac tone in his autobiography: "I sometimes think I regret that I could not have a group of infant farms where I could have brought up thirty pure-blooded Negroes on one, thirty 'pure'-blooded Anglo-Saxons on another, and thirty Chinese on a third—all under similar conditions. Some day it will be done, but by a younger man."[17] (The racial divisions are typical of the assumptions of Watson's generation about human genetic diversity: racial groupings were seen as the ideal controls for running an experiment on human conditioning from birth.)

Watson's wistful note is partly explained by the abrupt end of his scientific career. On Madison Avenue he was soon absorbed in bringing techniques of behavioral control to the marketplace. Modern advertising was just being born, and its ad men were only beginning to understand the extent to which people related to products not just rationally but emotionally. If one wanted to sell life insurance (or for that matter, hot dogs), one would do better to make a stirring appeal to the American way of life than to highlight the fine-print particulars of the policy (or sandwich meat) in question. Watson's experiments with white rats and babies taught him how to build and shape people's emotional responses to almost anything. Advertising was simply the vehicle for carrying this insight out of the laboratory, and Watson assured his new colleagues, "To make your consumer react, it is only necessary to confront him with either fundamental or conditioned emotional stimuli."[18]

Advertising's absorption of behaviorism through Watson was somewhat more complicated than his tendency to brag suggests. Appealing to the emotions in order to sway public opinion was not, of course, Watson's sole doing. The field's best professionals understood how to target irrational processes—that is, how to speak to that part of the human decision-maker that is not susceptible to well-laid-out arguments based on reason, particularly in the female sphere. As William

Esty, a J. Walter Thompson colleague of Watson's, wrote, "It is futile to try to appeal to masses of people on an intellectual or logical basis."[19] Eventually, working the "emotional appeal" and bypassing rational thinking became the basis of the territory mapped out by persuasive methods—as seen in advertising, public relations, polling, spin, and other techniques that register and modify attitudes held by masses of people.

Some historians and cultural commentators have exaggerated Watson's effect on advertising as if he were a Svengali, or assumed that he somehow brought a hard-nosed, exact science to bear on a primitive, inexact practice. In their view, he brought into vogue the "scientific sell"—using the authority of white-lab-coated types to enforce a product's appeal—which was to alternate with the "creative sell" for the rest of the century and on into the next. But Watson's influence was in fact subtler and, in a sense, more pervasive. It extended a scientific grasp into the domain of emotion and imagination. In an early version of human engineering, the complex of symbols and messages in an advertisement worked directly on such inchoate things as feelings, attitudes, tendencies, and preferences, changing them and indeed constructing them as people grew up interacting with the stimuli in their environment.

In popular publications, Watson continued to argue the case for behaviorism's special powers as a form of mind control, writing best-selling books putatively about baby care, the main goal of which, in the view of one recent critic, was actually "to intimidate, infuriate, and titillate mothers" so as to weaken maternal tendencies to dote excessively on their babies. (One chapter warned against "The Dangers of Too Much Mother Love.")[20] He came out against coddling and took part in a movement gathering steam in the 1920s and 1930s, funded in large part by the Rockefeller Foundation, to bring childrearing under the purview of experts.

BY THE MID-1920s AND 1930s, the "behavioral revolution" had spread from psychology into other social science fields, captivating many of the younger faculty, but it still had a bad-boy edge to it. Sociologists, political scientists, and economists who declared themselves behaviorists meant not that they had suddenly taken up rat research but rather that they were convinced of the central insight of behaviorism—that mind and matter are not separate, that there was no "black box" of consciousness or mysterious something that could not be explained, and that therefore a science of social control through behavior control was possible, indeed imminent. To be a behaviorist was to share a kind of attitude and at root a conviction. No longer an upstart upsetter of conventions, behaviorism became a sleek if suspect vehicle in which social scientists, especially psychologists, rode to solidify their place at universities and elsewhere. (Remember that as recently as 1900 psychology was in most places a mere subset of philosophy, hardly ever considered its own discipline; sociology and anthropology sometimes existed as separate departments, sometimes not, but in any case were not very powerful.) As young Turks who had succeeded in overthrowing an old order, behaviorists took some time to survey their winnings. And then, with the air and sometimes stridency of revolutionaries and cobweb-sweepers, they set about changing not only laboratory practices but the ordinary life outside. "Behaviorism called for new laboratories and even new words," wrote Watson.[21] The maze-running tradition had arrived.

CHAPTER 3

Embracing the Real

IN 1945 THE PRESIDENT of the University of Chicago, Robert Hutchins, went on record with a peculiar idea, or what may have seemed to some to be a sign of encroaching senility. In an interview with a national magazine, he suggested that "the founder of the social sciences in America" was a man named Beardsley Ruml. The reason, he went on to say, was that so many of this man's ideas had been implemented in the social science field.[1]

Of course, President Hutchins's opinion may not sound particularly tendentious today, for really, the issue of who invented the American social sciences is not a hotly debated topic, on the order of who made the first pizza or thought up the model T, the ATM machine, or the shape of the football. Yet in terms of their impact on daily life, especially in this country, the social sciences have had an enormous effect. Out of these sciences have emerged many of the measuring and engineering techniques that American society, more than any other, has experimented with and adopted: advertising techniques, public relations strategies, therapy movements, propaganda campaigns, focus groups, emotional management devices, human resource sciences, "crunching" knowledge from data processing, data mining. These social-science-bred techniques take the measure of what is human, and in so doing they change it.

Still, even if one grants that the social sciences have been uniquely important in America, one may be less willing to admit this of Beardsley Ruml. Calling him the founder of the social sciences appears to be evidence if not of willful perversity, then certainly of silliness. Compare Ruml, for example, with nineteenth-century pioneers like Karl Marx and Auguste Comte and Herbert Spencer, or with Americans like William James and John Dewey, and you will have little to go on. He is perhaps best known for reforming the income tax. Before 1945 the Internal Revenue Service required payment of one's taxes at the end of the year, causing an annual national headache and heartache. Ruml came up with the idea of pay-as-you-go. He then topped off his career with a stint as chairman of the Federal Reserve Bank of New York. Along the way, he walked a strange path, completing a Ph.D. in psychology at the University of Chicago and eventually becoming an industrial psychologist, foundation director, university dean, corporate executive, head of Macy's, and banker.

But the founder of the social sciences in America? How so? No one, not even social scientists, reads Ruml, perhaps because he wrote no books. And rarely is he mentioned in the classroom. His name comes up only twice in Dorothy Ross's authoritative *The Origins of American Social Science.* Yet he was in his time almost universally acknowledged as a brilliant social scientist, one who excelled at "producing ideas in brilliant cascades." He possessed, according to Louis Brownlow, "one of the most complex and comprehensive minds of modern times."[2] Fulsome profiles appeared in the press, announcing Ruml's importance to the nation. On the strength of his gift for coming up with new ideas, some saw him entering public office or running for president. Even though he did neither, many of his ideas came to fruition.

The idea that Ruml was the founder of American social science, however outrageous, is an interesting one to defend. Between 1922 and 1929 an arm of the Rockefeller Foundation gave out almost $50 million toward the pursuit of the social sciences around the world, and it was Ruml who, from the age of twenty-six, was in charge of dispensing it. And he did more than dispense: he had a vision—a progressive engineering one—based on getting closer to the reality of social life so that one might rebuild it to better specifications. He was a super-administrator, in the sense of aid-giver, system-builder, planner, dreamer, talent scout, talker, and engineer. Understanding why, how, and by precisely what means Ruml founded American social science is a way of delineating a historical crossroads that has led to our own era. At this crossroads was the promise of social and human engineering and the shared feeling that the newly energized social sciences were the logical ones to fulfill it. They shared a sense of a possible "embrace of the real," in Susan Sontag's phrase, that would power these changes—for to embrace the real, as a social scientist, meant to come into more direct contact with a greater number of social facts and a greater variety of human phenomena than had ever been possible before, so that a science of society could begin to work not through models or metaphysics but through the conditions and circumstances of ordinary people's ordinary lives.[3]

During the 1920s the officers and case workers of the major foundations, various government bureaucrats, and freethinking and nonfreethinking young visionaries all came to that crossroads and found they had distinct opinions about which way to go. As the historian Oliver Zunz describes it, during this time "a new institutional matrix of research" was in the making that changed how social research was done and what could be done with it, producing altogether "a critical and influential configuration of ideas, structures, behaviors, policies, and prescriptions."[4] In short, an attitude was emerging that reality, any reality, in and of its essence was subject to change. Social scientists had the tools and training to bring about the proper kinds. Furthermore, in their shared changing-of-the-guard attitude, these men were more energetic than the usual bookish ruminator on social questions. Ruml *defined* the social sciences for the first time as a collective entity capable of social and human engineering.

The precocious Ruml came from a family of mixed Czech and

Mayflower stock (as it was usually put) who had made it as prosperous lawyers in the town of Cedar Rapids, Iowa, despite a Czech grandfather who was a laborer. Ruml's Bohemian side came out sartorially, and he was wont to dress his large frame in capacious embroidered shirts accompanied by yellow or pink corduroy trousers (outfits sometimes considered an impediment to Ruml's seeking higher office, although by the time of his postwar eminence he reserved them for wear outside the office). He went to Dartmouth as an undergraduate, where he worked with Walter Bingham, a pioneer in psychological testing of soldiers and businessmen, and to Chicago as a graduate student, studying intelligence classification of workers, which activities apparently cemented a native inclination toward mental agility and physical sloth. In lifestyle he seems to have made lavish use of Winston Churchill's dictum, "Never stand when you can sit and never sit if you can lie down," but he was also known as a bon vivant, a wine lover, and a charming conversationalist.

He was somewhat famous for his method of acquiring ideas, a method he advocated to others. This entailed basically doing nothing but sitting at his desk and waiting in "a state of dispersed attention"— a quasi-meditative disposition akin perhaps to John Keats's famous "negative capability" or the practice of loosing oneself from external awareness that Aldous Huxley called "deep reflection." The result was the ability to see things anew, without the usual preconceptions.

Even in the realm of seeing things anew, Ruml was notably practical, for his ideas were always about reality and how social science could better approach it. Along with Walter Lippmann, the journalist-booster of science-applied-to-democracy; Elton Mayo, the father of industrial management and human relations; Harold Lasswell, an inventor of propaganda studies; and various members of the famed Chicago School, he shared a common sense that Americans were on their way to a place of new promise where culture and society and human desires could be reenvisioned and remade. "We stand on the threshold of a new era," announced Ruml's friend and colleague Robert Yerkes, when "human engineering will shortly take its place among the important forms of practical endeavor."[5] These thinkers can be seen as straitlaced counterpoints to the surrealists, juxtaposing unlike things and ignoring accepted categories, making the strange familiar and the familiar altogether surprising. But unlike many artists and social skeptics of the day, these men were on the side of a brisker science and styled themselves as designers of human social life—and of human beings.

RUML is a convenient starting place to look at the invention of a modern American social science, which could not have taken place without the concomitant rise of modern-day foundations and the goals that powered them. First one must consider the post–Civil War money pool. In 1880 there were fewer than one hundred millionaires in the United States, whereas in 1916 there were more than forty thousand. Twenty men had accumulated many millions each, and one of the wealthiest, John D. Rockefeller, Sr. (America's first billionaire), founded the biggest foundation.

Even when he was a poor man, Rockefeller always observed biblical tithing, keeping an exact record of his ten percent in neat notebooks he called Ledger A. In 1859, for example, he gave $72.22 to help a Cincinnati freedman buy his slave wife. Once Rockefeller had become very, very rich through, in his words, "the difficult art of getting," he set about the equally difficult art of giving away what eventually amounted to $600 million. He started his first organization for doing so in 1901, the General Medical Board. Its aim of fighting ravaging diseases around the world was possibly psychic reparation for Rockefeller's irascible father, Big Bill, a man with no medical training save the selling of (literally) snake oil, who used to travel around advertising himself baldly as "Dr. William A. Rockefeller the Celebrated Cancer Specialist."

Soon philanthropies such as Rockefeller's were springing up all over the United States. By 1926 there were about 150, accounting for nearly a billion dollars in capital. Most were devoted to specific causes or charities, and each foundation typically named a small body of trustees, organized itself legally as a corporation, and applied the earnings from the principal endowment toward whatever goals it decided on.

Why did newly rich men, these captains of industry, found foundations? The courts, after all, frowned on it. The jurisprudential view was that corporations should be encouraged neither to be donors nor to engage in philanthropy, and the public was suspicious, not expecting great benevolence from profiteers. Furthermore, men like Rockefeller were not in need of tax shelters, for there was no significant income tax on their immense profits. The spur, argues the historian Judith Sealander, was "a need for better organization" in society at large.[6] Rockefeller policy documents dramatize how the goals of making a better-organized society coalesced after 1922, gaining peak momentum due in large part to the plans of Beardsley Ruml. Reading through the foundation's evolving policy papers like a story, one can draw out this plot: When the trust began, its creators felt the need for general betterment of people's lots and for "social regulation"—that is, bringing order to the chaos of social life. Next, the social sciences made their debut as the meliorative agent, the key to knowledge that would bring about change. By the end, the trustees felt confident in the efficacy of such social science investigations, and with the means found, the goal of the foundation was solidified: "social understanding and social control in the public interest."[7] In this way, it was felt, democracy might be preserved, not through noblesse oblige but through science.

AMONG THE THOUSANDS OF FOUNDATIONS, five set out to shape public policy. The two biggest were the Carnegie Corporation, through which Andrew Carnegie, who did not believe in family inheritance, gave away most of his self-made $350 million, and the Rockefellers' related trusts, controlled by Rockefeller Sr. and Rockefeller Jr. On a smaller if still significant scale were the Commonwealth Fund of Standard Oil associate Edward Harkness; the Russell Sage Foundation, named by widowed benefactress Olivia Sage for her miser husband; and the Julius Rosenwald Fund, based on the Sears, Roebuck empire's fortunes. Established in 1936, the Ford Foundation made a late-coming sixth. Carnegie and Rockefeller, however, dwarfed the others in size of endowment. What distinguished these six was that, rather than targeting individual emergencies or personal tragedies or hard-luck stories, each stated a desire to "improve mankind." They saw themselves as expert engineers engaging in social experimentation and in an effort to stay flexible did not concern themselves with "private" problems except through mass-scale public programs.

Most foundation officers and staffers were of a certain type. All told, especially during the first thirty years, their world was made up of a few hundred people, mostly of East Coast WASP extraction, Ivy League educated, living in New York City, and more or less dedicated to the general goal of social "reform." People moved from philanthropy to business or universities and back. There were also more women and in higher spots at the foundations than in other spheres. These men and women set out to influence public policy and public programs, and as they usually had at their disposal more money (a lot more) than the government agencies with which they dealt, they had heft and, more often than not, got their way.

The opening of the foundations' archives to public inspection during the 1970s has resulted in a fierce debate among historians: Were their founders capitalist tools or public-minded Samaritans? One group of scholars, who have been accused of "conspiracy-theorist" views and the misapplication of Antonio Gramsci's theories of hegemony, see the foundations as examples of the energetic promotion of the hypocritical, specializing in programs that, in the name of benefiting the powerless, consistently advanced the interests of those who already held power. Commonly used phrases like "social technology" and "social control" tend to evoke this reading, as do particular cases. For example, Rockefeller Jr. was both president of the Rockefeller Foundation and a member of the board of directors of the Colorado Fuel and Iron Company, infamous for the 1914 Ludlow Massacre of striking workers and their families; meanwhile a major beneficiary of Rockefeller Jr.'s own personal money was Elton Mayo, whose program to adapt industrial workers to their tasks by deradicalizing them through psychological counseling would seem to suggest this emerging social science was anything but neutral. (In 1917 Rockefeller Jr. himself acknowledged "the fear which many people have of this great fund."[8])

Others see in the foundations some decent folks who only wanted the greatest good for the greatest number, and if their programs sometimes went awry or contributed to greater problems than the ones they were attempting to solve, it was in the nature of such an ambitious undertaking. The staff members and trustees of the foundations were largely free to act independently of oversight, but they had no dire or secret plan to wrest democratic freedom-of-action or freedom-of-thought from average Americans. (We will see, however, that this was at times the effect.) The staff and trustees believed that the outside imposition of policies and of normality itself was necessary for the smooth functioning of any social system, even (and perhaps especially) a democratic one. Democracy, in short, was frightening and, in the absence of heavy-handed authorities of old, might devolve into mob rule. Accordingly, a new science-based authority, streamlined and logical, would bolster the chances for true democracy to survive. Hence the foundation visionaries and the social scientists they supported longed to engineer social reality and the human beings who lived in it. In time, the embrace of the real came to mean for them a social system of control brought inescapably into the very corpuscles of the human organism.

THE ROUTE Beardsley Ruml took to unify the American social sciences—a unity based on a new embrace of reality—lay directly through Rockefeller's trusts. Ruml's first platform for his ideas was at the Laura Spelman Rockefeller Memorial, the last of seven philanthropies that had been made out of Rockefeller Sr.'s money.

In October 1918 Rockefeller Sr. had incorporated the Spelman Memorial in tribute to his recently deceased wife, and it was meant to support the causes she had espoused, such as Baptist missions, churches, women's and children's welfare, and homes for the aged. With an endowment of $74 million, the Spelman Memorial was in a position to do something substantial in these areas. Unlike Rockefeller

Sr. himself, whose father was a con man and a bigamist and whose long-suffering mother had little time for crusades, Laura Celestia Spelman had come from a long line of Puritan-stock activists, a "family of genuine substance."[9] Evangelical in their crusades against saloons and rum-drinking sinners, they also promoted public education and the abolition of slavery and made their home a stop on the Underground Railroad. During her long marriage to Rockefeller, "Cettie," who in her early years was an ardent feminist and who railed against the "almighty dollar" in her college writings, devoted herself to spartan living and welfare projects. At her death, married though she was to the richest man in America, her closet revealed that she had only ten hats, worth ten dollars.

In Cettie's spirit, the memorial's officers initially vowed to give immediate succor to practical causes. They would not contribute to academic theories and philosophical musings about mankind but would instead show "a practical interest in the welfare of individual men, women and children." The Young Men's Christian Association and Young Women's Christian Association were two of its main beneficiaries. By the early 1920s, however, when the memorial had not been in existence but a year or two, its own executive committee began to waffle and shortly resolved to end all support for charitable contributions, immediate reform, or "direct social welfare."[10] They were responding to a paradox inherent in trying to make a better society: the officers felt that serving "immediate utility" and sad plights was a direct way of using their money, but of what use was direct when it ultimately served only indirection? As in the riddle "At what penny are you rich?" benefit bestowed individual by individual or penny by penny would never add up to a society-wide good. The old philanthropic model was obsolete, for it was clear that beseeching letters from poor families could hardly be answered one by one. "Society" was a quality that could not be approached through the quantity of needy persons who existed each in his or her unique need. Benefit on a society-wide scale could happen only through programs based on scientific study. By a logic common in those years, the very existence of society was firmly linked with the very operations of science, and the foundations of the great capitalists arose to forge that link in a new way. The memorial's board began casting about for a full-time director who could move in this new direction. The YWCA's secretary was considered for the post but deemed too reminiscent of the social services.

In 1922 Beardsley Ruml stepped in to head the memorial, at first only as interim director due to his youth. But he soon won unusual powers, secured his place there through the remaining years of its existence, and developed a long-range program that favored the social sciences. By this time he had already served as assistant to the president of the Carnegie Corporation, James Angell (his graduate school adviser), and had developed psychological testing for the War Department, also known as the "classification of personnel," under Walter Bingham (another of his professors). The war experience inspired Ruml in 1919 to found the Scott Company, consultants and engineers in "industrial psychology," along with Bingham and Walter Dill Scott. Based on scientific management principles and F. W. Taylor's streamlining-of-behavior methods, industrial psychology was a way of fitting workers to their jobs (and jobs to their workers) by means of "mental engineering," "psycho-technology," and workplace efficiency gauges, thus inaugurating a fusion of psychology's tests and measures with corporations' labor needs that continues into this century.

The Rockefeller job went to Ruml as the result of a confluence of factors. For one thing, Angell, who was by then president of Yale, lobbied hard for him, convincing the foundation's president, Raymond Fosdick, already a fan of Ruml's and a confidant of Rockefeller Jr.'s, to press his cause. For another, important and connected people in the ambitious field of psychological adjustment—such as Charles Merriam of Chicago and Robert Yerkes of Yale—already considered Ruml one of the "leading men" of his generation. Within a short time his spot at the memorial was secure, and its policy statements from the 1920s became his wish list. That wish list had a remarkable record of being realized, for Ruml not only had plenty of chutzpah but also plenty of follow-through. As he told his colleagues in 1927, he organized the memorial so that it would operate as an organism, not as a bureaucratic machine.

Ruml kept it highly centralized yet informal. One of his major administrative innovations was to stress giving out large "block grants" over three to five years rather than small individual ones. A few leading universities received huge grants to build up leading-edge social sciences, which marked a modern phase in the field and took it beyond the lone scholar working in a library, supervising a few Ph.D. students, and favoring impressionistic analysis in a kind of vacuum. Ruml's new strategy hinged on these administrative innovations, which meant that in effect the substance of his program was undivorceable from its delivery: not only ideas but how one promoted and sustained them mattered, especially when the aim was to effect significant social change. Group projects, giant in scale and sweep, began to emerge.

Having a large amount of money to disburse, much of it not committed in any way, Ruml brilliantly appeased the old guard by arguing that the old aims of Social Welfare and the Betterment of Mankind could be best fulfilled through the social sciences. First he defined them, something that had not been done before: social sciences included sociology, ethnology, anthropology, psychology, and certain aspects of economics, history, political science, and biology—that is, any field that contributed "a body of substantiated and widely accepted generalizations" about human capacities and behavior.[11] Defining a new field by bringing together many existing ones that shared a common goal was a stroke of genius, for now social science in its very definition brought about change in social life through human engineering. There had previously been hardly any support for these fields, and suddenly there was a great deal.

The memorial's policy papers from these years (1922–26) used the term "Social Welfare" pretty much synonymously with "Social Engineering" and "Social Intelligence" and "Social Technology." Once the new social science had explored and mapped the human and social realm in a properly replicable and as-objective-as-possible manner, change would necessarily follow. Under the banner of scientific reason, even the irrational elements of society were susceptible to control. Crime, delinquency, and abnormal sexual or familial function could be corrected through the redesign of environmental and social situations; perhaps scientists could even address unbelief and the ravages of twentieth-century normlessness. Controlling social technology, it followed logically, was a task for social scientists and other experts, for who was in a better position to render knowledge as technique? Faced with the monumentality of their task, human and social engineers reminded themselves that they were mere servants or technocrats working for those who set forth the ultimate goals of social control: democratically elected officials. Without social-control experts, such leaders, short of adopting authoritarian methods, could hope for little effect. Thus social sciences were the greatest hope for democratic social control.

The melding of science and society, social control and social engineering, according to the Ruml-led memorial, should ideally take place in a Social Laboratory, the implausibility of whose actual existence scientists and foundation officers often rued. The subject matter of the social sciences by its very nature was resistant to laboratory study, they believed, for it "is extraordinarily difficult to deal with," said Ruml in a key memorandum in 1922. "It cannot be brought into the laboratory for study; elemental phases are almost impossible to isolate; important forces cannot be controlled and experimented with, but must be observed, when and as operative."[12] The social sciences faced a stumbling block, he continued: they could not be like the physical sciences because "the hypotheses of social science can only rarely, if ever, be proved by laboratory methods." Then the rub: "Consequently, *the possibilities of social experimentation are to be kept constantly in mind*; and opportunities for practical demonstrations are to be utilized whenever they promise to throw light upon the validity of tentative social findings."[13] So it was that this new approach (social science at large) was baptized in the waters of the real (experimentation and practical demonstration).

A theme running through Ruml's documents is the reenvisioning of American can-do pragmatism, made somehow more pragmatic. In his 1929 address on the social sciences, Ruml could speak in triumph of a "new objectivity" and a "new confidence," and he congratulated his listeners for sloughing off old worries about whether they were adequately quantitative and letting epistemological dialectics of the why-are-we-here, where-are-we-going variety go unworried over. As a result, he said, the social scientist was able to seek out *realistic contact with the raw data of his problem.*" Ruml and others had a strong impulse to come in contact with reality itself, raw, concrete, red in tooth and claw, or otherwise. Seeking it was a sort of hunger, as Ruml pointed out:

> In order to secure a background comparable with that given in natural science, regardless of later specialization, the student would require contact with the slum, with the Gold Coast, with Bohemia, with the laborer, small merchant and farmer. He would need direct experiences with foreign offices, with boards of directors of large and small corporations, with political committees, with trade unions. He would require an understanding of personality disorders, of the working of primitive as well as of advanced societies based on personal participation.[14]

Even in his earliest policy document Ruml argued that "means must be devised for securing a far more intimate contact of the social scientist in the university with concrete social phenomena." This talk of "intimacy," of "direct contact," of the lure of "raw data" sounded in some ways like Walt Whitman lusting to embrace reality in *Leaves of Grass*: "I am mad for it to be in contact with me." Ruml and his group harbored a similar push for contact. Scientists felt a quintessential longing for the touch and feel of real things, even though to touch and to be touched meant still maintaining one's hard-won distance. One could not, *à la* Whitman, go down to the bank by the wood and become undisguised and naked. As the social sciences emerged from the nineteenth century, their aim was to move toward an intimate encounter with life itself in all its variety, while still keeping a cold observer's eye on it. An ambiguity remained: that which one contacted—the reality, the direct experience—was still a specimen, an object, an other viewed in a laboratory (metaphorical or real) under controlled conditions. Intimacy had its limits. The more social scientists embraced the real, the farther away they needed to remain for the purposes of accuracy and control. Penetrating the object of research meant maintaining a paradoxical distance.

Encountering the raw stuff of reality and then taking that experience into a laboratory in order to build models and scientific procedures for controlling and therefore bettering social life was the plan Ruml and others advanced. In these years it took the form, as one sociologist put it, of a "grass-roots empiricism," which despite its populist-sounding rallying cries ("grass-roots," "facts," "the real," "direct") was linked to scientific positivism and pitched in battle against the humanism of earlier approaches.[15] The watchwords were "order," "control," and "objectivity." Initially these words were more than a matter of sloganeering, brute imposition, or naïve scientism, even if all these things existed and would come more or less into play as the years progressed. Ruml's plan was to try to tap the wellspring of reality itself and bring it into a controlled environment, the laboratory, not through a fleeting embrace but through rigorous observation and the capture of real things.

THE FIRST PLACE where Ruml's plan was fully accepted was the University of Chicago, which was the biggest single recipient of Ruml's memorial money (at least until the Yale Institute of Human Relations came along in 1929 to win an unprecedented "windfall").[16] During the 1920s Chicago's total bequests of $3,389,000 were far beyond amounts going to other key beneficiaries, namely the Brookings Institution, Columbia University, the London School of Economics, and Harvard. This made perfect sense, for ever since its founding in 1892 with Rockefeller Sr.'s money as "Mr. Rockefeller's university," the Chicago approach had distinguished itself by the way it looked at its subjects close up (the city's paupers, cast-offs, Polish peasants, almshouse dwellers, wine room celebrants, and juvenile delinquents). From the start, Chicago social scientists sought to make contact with the "real" in their amazing city. Or perhaps it was the other way around, for Chicago had a way of coming at you, as John Dewey observed: "Every conceivable thing solicits you; . . . things . . . simply stick themselves at you, instead of leaving you to think about them."[17] There was a strong tradition of reform-mindedness among the faculty, who were also members of various special parks commissions, civic federations, and settlement houses.

Chicagoans pioneered an engineering approach writ large, buckling the real to experience and experience to change. As we have seen, Jacques Loeb worked there on his tropisms, monstrous hydras, and artificially propagated sea urchins. John Watson trained there and inspired animal researchers in laboratories across America to base their work on the same action-philosophy. In the 1920s Chicago was in full flower: a unique dedication took hold to observe particulars and to analyze them in relation to the whole. Famous in this regard are the anthropologist W. I. Thomas (a dandy, cosmopolitan, freethinker, man in search of experience) and the sociologist Robert Park (a newspaperman, allied with Booker T. Washington, who taught innovative classes on "The Crowd and the Public," "The Newspaper," "The Negro," and "Methods of Social Research"). W. I. Thomas and Florian Znaniecki's *The Polish Peasant in Europe and America* (1918–20) broke new ground and turned an ethnographic eye on people usually out of its range.

The many programs to which Ruml funneled money and support shared something of this encompassing approach to reality and the accompanying quest to establish laboratories for studying it. But in the name of real data and real experimentation, they also ushered in an era in university research in which the "grant swingers" with major foundation contacts dwarfed individual researchers, so that, as Harold Laski wrote, "the man who dominates the field is the man who knows how to 'run' committees and conferences, who has influence with, and access to, a trustee here and a director there."[18] Getting cooperative research going among experts across fields took some organizing and administering.

For some reason, the advent of startling new social science paradigms in twentieth-century America was often accompanied by campaigns of "fundamental" research on babies. (Perhaps this was because babies are evidently so fundamental and yet so mutable.) Watson had

done it, and starting in 1923, the memorial stepped up its support for
the New Psychology and an effort called the child development movement. Ruml's key associate here was Lawrence K. Frank, an economist
from Columbia only four years older than Ruml, who was described by
one of his beneficiaries as "the procreative Johnny Appleseed of the social sciences, a peripatetic horn of plenty crammed to his lips with
everything that's new, budding, possible, and propitious."[19] Sounding
like a less contemplative and less sedentary Ruml, Frank also on occasion received credit for founding the modern behavioral sciences (for
example, from Margaret Mead).

Frank and Ruml sponsored experimental classes in parental education, to introduce methods of positive conditioning and formation of
good habits—bringing behaviorism into the home. The bibles of the
movement were Watson's *Psychological Care of Infant and Child* (1928)
and Boston psychiatrist Douglas Thom's *Everyday Problems of the
Everyday Child* (1927). Nurseries-cum-laboratories and laboratories-
cum-nurseries sprang up at the Institute of Child Welfare Research at
Teachers College in New York in 1924, the Psycho-Clinic run by
Arnold Gessell at Yale, the Iowa Child Welfare Station, and additional
institutes at Berkeley, Toronto, and Minnesota, with Harvard, Antioch,
and Columbia Medical School following their lead in the 1930s. A re-
demptive science of childrearing was the goal, based on the increasingly self-evident premise that while it surely was a good thing to be
unlocked from lockstep social formalities and cultural patterns, just
drifting along without devising new standards to replace the old was a
waste of opportunity, one that ill befit science as well as society. Social
and moral drift was to be replaced by control mechanisms that replaced tradition with order and purpose.

AS SOCIAL ENGINEERING GAINED MOMENTUM, ideas were
married to action in laboratory or laboratory-like conditions. One example is the Ruml-sponsored Hawthorne experiments. In 1926 Elton
Mayo was plucked from obscurity to preside, two years later, over one
of the most famous and influential social science experiments ever

conducted and to pioneer social experimentation in a quasi-laboratory. The child of an upper-middle-class family from Adelaide, South Australia, his grandfather the leading surgeon in the colony, Mayo had regularly disappointed his parents' expectations. Sometime later, in 1903, still at home in Australia and having engaged with philosophy to no great effect, he turned to the problems of industry and how men and machines interact. Finally he gained a post as professor of philosophy and psychology, working on a synthesis of fields (he "liked to present himself as an interdisciplinary renegade"), but he found the outpost university atmosphere stifling.[20] Like his friend the anthropologist Bronislaw Malinowski, he went farther afield. In 1922 Mayo arrived in the United States at the age of forty-two with only fifty pounds to his name. Only four years later he gained a position at Harvard and became one of the most esteemed social scientists in the country, bespeaking both his personal charm, to which many attest, and the goingness of the attitude he shared with Ruml and others.

What captured his interest was the question of the social radical or renegade and how his disruptive behavior could be controlled. Early in his career he studied leftist activists, revolutionaries, and rabble-rousers and understood their activities as pathologies that stemmed from imbalances in their bodily secretions, past histories, and minds. Once the pathology was "set aright," Mayo found, the activism would go away. A pivotal case study for Mayo was a thirty-year-old man, described in Mayo's "The Mind of the Agitator," who simply could not take orders from any authority figure. At each job the foreman, boss, or manager at some point so enraged the man that he took violent umbrage and inevitably lost his job. In terms of politics, too, the man was anti-authoritarian and upheld social revolution as the only possible cure for social inequality and injustice. Mayo found a cause for this behavior (his father beat him as a child) and a solution to it (talking things out through "nondirective counseling"). With the old pattern short-circuited and the revolutionary neutralized, the man could now work in peace, no longer troubled by left-wing ideas. This story of agitation, treatment, and subsequent middle-of-the-road politics was put to great use throughout Mayo's career, and it appeared in many of his writings with an import close to allegorical.

When he arrived in the United States in 1922, Mayo was desperate for work and contacted the Rockefeller-funded National Research Council–PRF, made up of interdisciplinary social scientists. They were intrigued by his ideas about strike control. He reported in a letter home to his wife that " 'a representative of one of the major foundations' is anxious to meet me and talk things over in New York."[21] The representative turned out to be Beardsley Ruml, and the meeting went well—the two men found they shared much in the way of family background and current interest in a science of society as well as a love of good food and wine. They began a close friendship, and Ruml came to serve as "the man largely responsible for Mayo's being so well-established in America's academic life."[22] In 1923 Ruml secured a spot for Mayo in the department of industrial research at the Wharton School and obtained a small grant for him to conduct industrial research there despite concerns voiced by the Spelman Memorial board, which was not yet fully behind Ruml and which found Mayo's work too overtly political. Since the 1914 Ludlow Massacre, however, Rockefeller Jr. had had a personal interest in how industrial relations could be made to run more smoothly, and so approached Ruml, offering to pay for Mayo's work personally. After Ruml went to Pennsylvania to observe Mayo's early experiments, he wrote, "I am very favourably impressed, not only by the quality of the work Mayo has done, but by the way he has interested manufacturers in the possibilities."[23] In September 1926 Mayo moved to Harvard and shortly began the Hawthorne experiments, in which he used the factory setting as a kind of laboratory and created a model for bringing social engineering into social life.

In discussing the Hawthorne experiments, it is useful to begin with a long view: for fifty years they have been evaluated and reevaluated, serving as a landmark in social science and being called the "first major social science experiment."[24] Reaching their peak of influence in the 1940s and 1950s, these experiments epitomized a shared understanding among scientists, business owners, and management, and set forth a radical new style of organizing and supervising industrial workers, especially those in repetitive assembly-line jobs. They were widely credited with putting the "human factor" back in the industrial equation (where scientific management had factored it out) via "human relations." Industrial sociology and personnel management followed in their wake and with their imprimatur. Within these fields and the related social sciences, the Hawthorne experiments "have acquired the status of a creation myth," writes their primary historian, Richard Gillespie.[25]

The usual view is that these experiments produced the kind of startling insight—a great "aha!"—that often attends the revelation of an obvious truth in a new form. What they taught was simple yet revolutionary. Scientists began testing factory workers for the effects of lighting on their productivity, expecting output to suffer as the lights progressively dimmed. Instead, they found that productivity either went up or remained the same. Mystified, they went on to vary other factors, and in each case (low humidity or high, many breaks or few, shorter hours or the same long ones) productivity went up. They attributed rising productivity to the fact that they were monitoring and paying attention to the workers. This produced what textbooks call "the Hawthorne effect": the experimenters were unwittingly influencing the experiment. But it also suggested a corollary truth. If management did likewise—entered into the work situation, paid attention to workers' psychological states, and even gave them therapeutic counseling—there would be less agitation and rebellion, especially through unions. This insight signaled a shift from F. W. Taylor's portrayal of workers as material to be molded in the most efficient manner. Now the human element was the focus of attention.

The most famous of the Hawthorne experiments took place from 1927 to 1932 in the relay assembly test room at the Hawthorne Works factory of the Bell System's Western Electric Company. The work of assembling relay parts was "unskilled," highly repetitive, and tiring yet required considerable dexterity: in putting together a single R-1498 relay, for example, the worker (as a filmed motion analysis showed) performed thirty-two separate operations with each hand. This work was usually done by females, almost all in their late teens or early twenties, who were typically Polish, Norwegian, Italian, or Bohemian immigrants. (The job was popular among girls living at home, not yet married.) An early team of researchers looked at the women's private lives, their home environments, and their social experiences to see how their attitudes correlated with job performance. Specially designed equipment measured productivity automatically, by means of a paper tape that moved through a machine. For each relay assembled and dropped down a chute, a hole was punched in the paper.

In 1928, after a luncheon speech he gave on "What Psychology Can Do for Industry in the Next Ten Years" had caught the attention of Hawthorne management, Elton Mayo was called in from the Harvard Business School to begin another phase of experimentation. He pushed the focus to worker psychodynamics. In an attempt to understand the "we feeling" in the modern factory, Mayo and his staff conducted twenty thousand interviews of workers and found that it was not the substance of the interviews but the very fact of being interviewed (i.e., being given attention, scrutinized, surveyed, watched, listened to) that defused their griping and made them more docile and "better adjusted."

Although no absolute conclusion can be drawn from these experiments, they generated a huge mass of data, whose interpretation researchers skewed to the expectations, philosophies, aims, ambitions, preexisting views, and post hoc attitudes of Harvard social scientists, in-house Western Electric researchers, Rockefeller Foundation officers, and others. The data itself was shaped even as it was obtained. Thus controversies over the meaning of the experiments were not sidelights to some revealed nugget of truth but "an intrinsic part of the experiments themselves"; even the design had a "political character" and was the fruit of widespread worries about the growing strength of unions.[26]

Mayo's aims, political and scientific alike, were based on an "adjustment" paradigm that prevailed among his forward-thinking circle. The very goal of adjustment assumed that people and things work together as an organic system, and that when either things or people are out of whack, the system's functioning will suffer. Therefore the task of the "functionalist" (as some called themselves) was to adjust the one to the other. For Mayo, this meant that the work environment could be scientifically calibrated to bring about maximum adjustment of the worker to his role within the industrial process. A greater cohesion, a subtle solidarity, and a strong sense of "belongingness" among workers would benefit the system itself. This process might have sounded ominous (for in the name of benefiting the abstract "system" and "social order," it seemed mostly to benefit the entirely concrete factory owner) except that social scientists were very careful to explain that, as adjusters, they would be entirely neutral—they had no axes to grind, no prejudices to foist. They would be "neutralist technicians," as William Whyte put it.[27] Meanwhile workers were encouraged to adjust and adapt to their positions in so subtle a manner that they would hardly realize it. Overall the Hawthorne experiments betokened what was fast becoming a general goal for social science: to be so adept at engineering the situation of human beings that the engineered person would scarcely know it had happened.

IN THIS WAY, through countless projects sharing this attitude and apparatus, Beardsley Ruml "invented" the American social sciences and saw them professionalized, institutionalized, bureaucratized. (Meanwhile Ruml eventually cut his ties with academia and went on to serve on many practical-minded boards, among them the Market Research Corporation and the National Planning Association.) The promise of an embrace of the real—as fed by Ruml's program, Rockefeller's money, and the dreams and hopes of many—ended perhaps paradoxically in a rather narrow view of reality. Eventually the more generous impulses of social science to connect, to make contact with, to juxtapose, to see more clearly, and to invent—to "treat life not as something given but as something to be shaped"—were largely transformed into their opposite: to winnow down, to prevent, and to build systems of control, adjustment, and persuasion, escape from which would be ever more unlikely.[28] Ruml and his cohort started out seeking intimate contact with the poor, the suffering masses, and the dispossessed but ended in putting people at a distance as mere human material. And then, just as Ruml was leaving the memorial and the memorial was dissolving into the foundation, just as the stock market was crashing and a new impecuniousness was about to take hold, the hugest grant of all came through, bringing into being the Yale Institute of Human Relations, which dwarfed anything that came before or since. Via this unlikely leviathan institute, the Rumlite brand of American social science continued on, with its ineluctable impulse toward human engineering and social control.

PART TWO

Rooms: Freud and Behaviorism Come Together

CHAPTER 4

Psychic Machines

IN MIDSUMMER 1939 the president of the Rockefeller Foundation received a coy but serious invitation. A Yale sociologist had written to share his excitement over recent events at the laboratories of the Institute of Human Relations. "We have a rat up here who learns under most interesting circumstances," the professor explained, and he wanted the president to visit and see the rat post-haste.[1] So it was that the head of the largest foundation in America, itself founded by the richest man in America, went up to New Haven to watch an albino rat navigate a laboratory maze.

With a $7.5 million initial grant and additional monies eventually totaling over $12 million (equivalent to around half a billion in today's dollars), the institute remains the most lavish and ambitious social science project ever undertaken. "Here," offered *The New York Times* on the occasion of the institute's founding in 1929, "is to be brought into a compendium of wisdom all that the varied natural and social sciences have learned about this adventurous piece with which Nature has crowned her creation—a cosmography of the human being."[2]

These extravagant words befit a project that set out to use science to cap history's accumulated knowledge concerning human beings and their place in the world. Many invoked Alexander Pope's lines, "Know then thyself, presume not God to scan; / The proper study of mankind is man," cementing a connection to the Enlightenment project of gathering perfectible knowledge of mankind (but not preventing one wry newspaper type from commenting, "Pope wrote it, Rockefeller financed it, and Yale believed it").

During the first third of the century, foundations, social scientists, and other progressives had cultivated social and human engineering plans on an unprecedented scale, making the Yale project far from unforeseeable or unheralded. Yet the sheer size of what its recipients called the "windfall" was a shock to many, and its timing, coming as it did a couple of weeks before the stock market's Black Tuesday plummet, verged on the macabre. Just as Yale set to work with New York architect Grosvenor Atterbury planning a stately Georgian-style, million-dollar building to house its institute, the nation's finances collapsed. Still, the main players were optimistic. Their collective aim would be to "carry on research upon the basic problems of human nature and the social order" and house a vast laboratory for understanding humanity and society.[3] They would seek solutions for juvenile delinquency, mental illness, labor unrest, crime, and family breakdown and unify all the social sciences bearing on these problems. In 1932, as the crisis lengthened and four million more workers lost their jobs, Yale's institute continued to encourage the hope of alleviating the nation's suffering through its unprecedented program.

Coinciding as they did with the onset of the Depression, the institute's high-minded aims created a distressing sense of obligation and unreality. Not surprisingly, it took some time to get off the ground. To the chagrin of its funders, scientists and scholars from different departments—numbering anywhere from 21 to 128, according to different counts—operated under its "umbrella," and the approach tendered by its putative head, Mark May, was rather vague. May favored a touchy-feely-ish program to study how people attained certain personality types in different cultures. In one instance, he dispatched social investigators to working-class neighborhoods in New Haven to conduct research on twenty-nine families of problem youths. A field report described the case of "G.," an adolescent girl from an Italian immigrant family who had lost her virginity, was slack in dress, and frequently provoked family clashes. An institute social worker took an active role in the girl's life and an institute psychologist provided counseling. The subsequent report mentioned G.'s improved poise and personal appearance, her birthday parties held at the institute, and a crush she had on her Yale psychologist, followed by the oddly poignant note, "Institute paid for swimming lessons."[4]

Needless to say, this approach did not please the foundation, whose officers preferred an emphasis on behavior, especially a hard-hitting behaviorism that would reveal possibilities for engineering human activity and controlling it. More dismally, Rockefeller's general opinion after several years was that "no ideas of surpassing importance have come from [the institute]."[5] Unless the institute fulfilled its original mandate—a concerted "attack" on human behavior through the unification of the social sciences—and did so with alacrity, it would have to give the money back, a prospect that left the institute's director "much concerned." (Indeed, he was thinking of decamping for Hollywood, where he'd had a job offer to make motivational films.) With the head of Rockefeller's social science division "insist[ing] that Yale did not assume an impossible job," the institute spurred itself to action.

Those running the institute faced the task of making it something more than a cash cow for sustaining piecemeal projects, and of fostering a collective effort to advance a systematic program. In 1934 a little-known experimentalist named Clark Hull, a professor in the Yale psychology department, came forward to offer a research program that was already gathering adherents among the younger scientists. Here is how Hull laid it out:

General Nature of Program: 1) A thorough and systematic exploration of the conditioned reflex as the strategic place to locate the basic laws for a rational psychology. 2) A vigorous exploration of the high mental processes for evidence as to whether the conditioned-reflex principles are operating in them.[6]

From the outset, Hull was confident that he already had the answer to number two. His interest lay not in determining *whether* conditioned reflexes operated in all higher processes but in finding the practical implications of his absolute conviction that they did. Such confidence was not a drawback. Hull believed his program would help legitimize the social sciences and bring social life and human impulses into a new realm of calculability. For Hull, only radical change would do: emphasizing a systematic approach modeled on the natural sciences, and a mathematically based ambition to encompass all human phenomena, he presented his ultra-behaviorism as salvation.

Half a decade after the institute's much-heralded founding, with the public no longer interested and the press otherwise engaged, the social scientists there found their purpose. Toward the end of the 1930s they felt they were, for the first time in history, making a true science of humankind and laying down "the foundations of a basic science of human relations."[7] Psychologists as well as anthropologists, sociologists, and psychoanalysts joined together to craft experiments that mixed an easy-to-use, all-purpose version of Freud's theories with an expanded version of Pavlov's, all premised on an equation between the behaviors that rats displayed when running laboratory mazes and those that humans displayed when running social mazes. In 1937 six experimenters took over the day-to-day administration of the institute and its budget and diverted funds from older research projects to their own, to cover unanticipated expenses such as the care and feeding of throngs of albino rats.

Success finally came to the institute in a less cheer-inducing form than originally anticipated, for its program—focused as it was on running rats through mazes—did not seize the uninitiate's imagination; nor did it change people's everyday lives, at least not at first. However, it did galvanize professional social scientists around the country in a perhaps unprecedented manner. A true social laboratory had been built, and Rockefellerites felt the institute had "finally begun to assume the form originally contemplated."[8] In rodomontades a network of scientist-scholars ballyhooed their success, and similar programs spread into university laboratories across America. This was the moment when the foundation president was called to New Haven to witness what its money and support had yielded.

"Rat learning," along with the efforts of other types of animals, lay at the heart of the institute's hopes for a grand theory that would explain the full range of human behavior and make it predictable and thereby controllable. In the institute's rooms and halls rats were put through their paces in various seemly and unseemly ways. They ran down dark tunnels toward a light or a feed box, halted in blind alleyways, were trussed in tiny harnesses, performed while drugged or caffeinated, mated under less than ideal circumstances, and were electrically shocked into clawing and biting each other in simulated pugilistic frenzy. These scientists had come to believe that understanding the behavior of laboratory creatures was the key to understanding and modifying how humans acted.

Nothing a human did fell outside what a laboratory animal could be made to do. Yale scientists were not alone in this belief: all across America, to a tempo accelerating since World War I, work with rats was taking over the social and behavioral sciences. But the institute gave this work a great unifying jolt and a new emphasis. Scientists began to believe they could explain every human thing—from love to war to union organizing—with the help of experiments run on rats. Their science aimed to study not only external behavior but also internal states of mind, previously accessible only through a process of introspection. They longed to describe these states in a unified and perhaps even mathematical language. This approach would persist for several generations, take over much terrain, and have an immense influence. Today it seems hard to believe that scientists willingly modeled human society on rat behavior in such a bald-faced way and that careers and institutions rose and fell on the strength of such a program. Though the program may now appear ridiculous, that is in part because its premises are today thoroughly taken for granted.

Methods devised at Yale and spread through a nationwide network of scientists—methods for conditioning humans through cues and symbols, playing on their fears and desires, and creating, in the end, a laboratory version of the "self"—not only became the basis of a new science and a new technique but gradually filtered into life itself. Shopping malls and multiplex coffee shops, media spectacles and managed care outlets, the dark recesses of people's secret selves as well as the clean and well-lit social environments in which they dwelled, became experimental spaces for a new style of behaviorism. Combining chaos with predictability, the human factor with fact-based logic, and demands for freedom with a distinct desire for constraint, this untoward science became a part of reality within controlled spaces and in the fastnesses of the emotional life as well. Scientists were convinced that if they could control the behavior of albino rats—whom they had already trained to be "lever-pressers, chain-pullers, button-turners, or maze-solving experts"—they could also control human behavior. It would be a science of fitting "pegs" into "holes."[9]

THE CORE OF THE INSTITUTE'S PROGRAM lay in the work of Clark Leonard Hull, a man who, like John Broadus Watson before him, had come from undistinguished farmland roots and achieved an almost cultlike command over American psychology and the related social sciences (at Yale Hull used to call himself their "Führer") before vanishing almost completely.

Hull's passions and peculiarities stand out in his 1918 Ph.D. dissertation, where he set up a template for his life's work from which he never deviated. He set out to examine the most abstract human capacity of all—the ability to reason—and to consider its "functional and quantitative aspects."[10] Hull believed that previous scholars, when studying the highest human abilities, had failed to be truly scientific (that is, to render their results in a quantitative and replicable form), but he would not. Experiments must mimic actual conditions as much as possible, he urged with some vehemence, for "the more completely the conditions of the life process are duplicated in the experiment, the greater the probability that the results of the experiment will be true of the particular life process." For Hull, a good experiment was lifelike. Having set this standard rather boldly, Hull went on to violate it to an almost absurd degree. The experiments he conducted had as little resemblance to "conditions of the life process" as could be.

To wit, Hull ran his experiments in a small research room in the psychological laboratory at the University of Wisconsin, where one large window of northern exposure allowed light to fall on a table forty-two inches high and sixteen inches wide, illuminating the facade of his "Exposure Apparatus." With this homemade device Hull flashed a series of cards with Chinese characters on them in front of a series of participants, none of whom could read Chinese. Sitting on a high stool to one side of the table, Hull cranked the arm of the Exposure Apparatus and judged the human capacity to generalize commonalities from nonsensical symbols. People found the task easier when the series of characters ran from simple to complex rather than complex to simple, and having proved this to his satisfaction, Hull commended himself for having, in the two or three years it took him to conduct the experiment, "made as great a contribution to our knowledge" as "has resulted from all the enormous amount of work done . . . since that time" by all other researchers combined. He attributed this feat not to any success in mimicking the conditions of life itself, as he had stipulated one should, but rather to the opposite: "The point is that by the use of abstract experiments there is an enormous economy of time and energy in reaching preliminary conclusions as to where the truth may probably be found in the complex material of everyday life." The main ingredients of this early effort—an abstraction of abstraction, a passion for building "thinking" apparatuses, the blithe assertion of apparent self-contradictions, and an infelicity with experiment itself, all accompanied by sweeping claims of significance—permeated Hull's mature work as well, making his dominance of American psychology for thirty years a mystery over which many succeeding scholars and scientists have wondered.

AFTER COMPLETING HIS DISSERTATION, Hull spent several years teaching at the University of Wisconsin, and in 1929, on the strength of his work on reasoning, Yale's new institute used some of its lavish funds to import him to the East Coast and set him up as a psychology professor. Five years of vacillation followed, during which time, as the nation slid further into depression and disarray, the Rockefeller beneficiaries fumbled about trying in various ways to achieve world-altering significance and somehow justify their good fortune. During this time Hull worked in a personal wilderness of his own. He composed a Byzantine yet strangely systematic program that he articulated in a series of articles published between 1929 and 1934. These articles had a galvanizing effect on the institute. They put forward a system of explanation that would account for any action or thought taking place anywhere in the world. Thoughts were nothing but actions, Hull argued, and therefore were susceptible to being changed or engineered in different ways. The articles were a prelude to revolution.

Provisional and cautious throughout (each hypothesis, as Hull advanced it, still required experiments to bear it out), he nonetheless boldly claimed to have arrived at nothing less than a "purely physical theory of knowledge."[11] A purely physical theory of knowledge had been the holy grail of philosophers and proto–artificial intelligencers for a long time—Plato's nephew Speusippus (407–339 B.C.E.) had tried to collect all human knowledge into one volume; Ramón Lull created the Ars Magna, a thirteenth-century machine for discerning truth by "bringing reason to bear on all things"; and Leibniz in 1673 envisioned a universal calculus of reasoning to decide arguments mechanically. Yet here was Hull, an upstart psychologist, claiming to be in possession of that grail. If knowledge, the apex of human activities, lived in the brain's synapses, then surely all variety of other behavior did as well. In 1931 he announced he had found certain "pure-stimulus acts" that "perform the enormously important functions ordinarily attributed to ideas. . . . While indubitably physical, they occupy at the same time the very citadel of the mental."[12] The physical and the mental were one and the same. Hull felt he had finally "explained" such things as purpose, success, contingency, choice, conformity, anticipation, inhibition, punishment, sophistication, and freedom—all through the stimulus-response trial-and-error experiments he carried out (or to be exact, would soon carry out) with rats in mazes. Armed with this—"the grandest of the grand learning theories," in the words of one historian[13]—Hull became at his peak in the mid-1930s, a decade after Watson and a generation before Skinner, the most powerful behaviorist in the world. But unlike these two men, each of whom served as the public face of his discipline and neither of whom would have been out of place in a *New Yorker* profile, Hull's eminence was highly specialized, confined to the awareness of the practicing social scientists in his own and related fields. Only within this group was his influence unparalleled.

His program enjoyed what two historians have described as "near hegemony in the heyday of neobehaviorism," a fact that, from their vantage point, given its manifest weaknesses, was "truly amazing."[14] And as another historian has mused, "Given the power and comprehensiveness of [alternative behaviorist theories influenced by Gestalt psychology] . . . it is very difficult to explain the eventual triumph of Hullian theory."[15] A third account of Hull's eminence likewise displays great discomfort: "Hull's system represents the ultimate attempt at establishing the psychologist's advantage once and for all [over his subjects]. . . . That his system is now a curious museum piece from which only negative lessons can be drawn could hardly have been predicted at the time."[16] This convergence of scholarly views makes one suspect that head-scratching over Hull has become something of a set piece among historians of psychology. But to represent his system as purely negative, entirely bad, or simply embarrassing is to promote a wishful view of what social science in the twentieth century could or should have been rather than what it was. A closer look at Hull's system in its context makes its triumph seem less mysterious and shows why it was so powerful.

In the course of his 1929–34 journal articles, Hull presented an encyclopedia of "primitive mechanisms," as he called them, that operated in the gap between stimulus and response. Like tiny workhorses, they carried impulses unerringly from stimulus to response. As pure vehicles of function, they operated automatically according to the built-in impulses of the organism. Among them were the conditioned inhibition (mentioned in 1929), the automatic trial-and-error mechanism

(1929, 1930), the subjective parallel (1930), the anticipatory defense reaction (1930), the short-circuit (1930), the purpose mechanism (1930), the anticipatory goal reaction (1931), the principle of redinte-gration (1931), the spontaneous variability of reaction (1932), the habit-family hierarchies (1932), the automatic transfer of effects (1934), the principle of irradiation or generalization (1929, 1934), and the associative tendencies of divergent and convergent types (1934). Hull depicted these mechanisms as operating through tiny threads of stimulus-response reactions woven between "WORLD" and "ORGANISM," and from 1929 to 1934 they grew ever more intricate, as if a spider were learning to spin better and more effective webs. As Hull argued, each minute mechanism helped to "adjust" or mediate the organism in its relation to the world. Some were "primary tendencies," and some were "corrective tendencies" that served to counterbalance the bur-geoning effects of others. Indeed, they were creating the inner life of an organism from outer events. The result was an elaborate map of how, precisely, thought is made up of conditioned reactions. Like many bio-logical functions, Hull's functions could also be recombined. Building blocks for complex mammalian behavior, they were themselves built of "certain combinations of more basic principles."[17]

The fact that human functioning could be broken down into simple if complexly interrelated mechanisms meant it could also be built up from these same mechanisms. And therein lay the crux of the matter, wrote Hull: it should be "a matter of no great difficulty" to "construct parallel inanimate mechanisms, even from inorganic materials, which will genuinely manifest the qualities of intelligence, insight and pur-pose." Thinking itself, he believed, could be built, and humans, even in their highest functions, could be engineered to reflect outside influ-ences. Hull suggested that his mechanisms were a form of pure con-nection, varying and yet the same, the oil of the machinery, the primitive corpuscles of the nascent machine or man. Anyone with the proper training could fix these mechanisms before his eyes, could see them right there, in the inarticulate "-" between "stimulus" and "re-sponse." Clark Hull's mission was to explain that "-", to make it speak, to elaborate it, and to found a science on it. No one had ever tried to do this before. The consequences of his project—which was later taken over by his followers at the institute and "merged" with the insights of Freudian psychoanalysis—are still being felt today.

Hull's core articles are notable for their strangely oblique language and untoward leaps of logic, studded by mathematical coalescences. At back is always the assurance that an immense complexity resides in even the simplest behaviors, for the play in these writings between simplicity and complexity is ongoing. Rat behavior is at once self-evidently simple and yet so obscure as to require the keenest skills of excavation for unearthing its buried meaning. Even the simplest act, such as a rat running toward a light when it hears a bell ring, is guided by an unwonted complexity. On the one hand, Hull asserts, these are simple reflex actions, but on the other, the physical and energetic patterns formed by their interaction are exceedingly complex. In fact, they are so complex as to be nearly invisible: "The details of the more complex action patterns are so concealed as to be almost impossible of observation."[18] They are elusive to the point of inscrutability, yet when isolated—and Hull's work tried to isolate them—they were revealed as straightforward and admirably simple, although at second glance there were always new layers of complexity. For Hull, the task of making an inescapably accurate science of human behavior was promising but appeared almost infinite in its challenges. And yet "ordinary observation" and simple accretion of facts always cleared the way. This tug-of-war between the plain-as-day and the hidden allowed Hull to jump from mere twitches to the highest human accomplishments, since if infinite complexity was housed in a tiny twitch, it could then logically qualify as a worthy basis for the higher activities of humans.

What stuff did Hull believe made up these simple-yet-complex "action patterns" that he had identified so painstakingly? Assuredly he believed they were "physical," as opposed to metaphysical, in a mathematical sense. Hull loved logarithms and seems to have had a fetish for mathematics, as one Rockefeller officer commented in a memo after meeting with him in 1937.[19] By the 1940s he had coined eighteen equations that accounted for animal and human behavior, and he expected more to come. He explained the advantage of his method this way: "This quantitative approach with the attendant possibilities of utilizing the potentialities of a metricized mathematics for the purpose of exploring the implications of its postulates is somewhat in contrast to the topological approach emphasized by Lewin"—a rival experimentalist—"with its seeming limitation to the qualitative."[20] In short, a mathematical approach had the advantage of allowing him to be, well, mathematical. Instead of locating his mechanisms of behavior in some exact part of the brain or body, Hull concentrated on the activity as it unfolded before him, transforming the barely visible into fine-spun webs of "metricized mathematics" and logarithmic curves. Everything could be rendered as a simple-yet-complex pattern.

WHAT WAS THE POINT of all this math? Some scholars have seen in it an attempt to render all aspects of human life as machinelike as possible. This might be called the "machine age" hypothesis. The historian of science Deborah Coon suggests that such tendencies to quantify, which she argues shaped the field of American psychology from the 1890s onward, were ways of "standardizing the subject" in tune with the demands of a machine age.[21] As new industries emerged, their need for an adaptable and compliant labor force coincided with the proliferation of social scientists who announced themselves as experts at "tuning" the population. But the urge to quantify was driven not simply by social scientists' need to serve capitalism better. In fact, Hull's real passion was for building thinking machines. Bridging the laboratory segment and the world segment, lurking in the footnotes of his articles and in the "-" between stimulus and response, were Hull's incipient machines.

As it happened, the idea of a thinking machine had first occurred to Hull years before, when he was still a Michigan farm boy. A near-fatal attack of typhoid from food poisoning left him with a permanently bad memory. Two years later, while working as a mining engineer at the Oliver iron mines in Hibbing, Minnesota, he contracted polio. While bedridden, he decided to study psychology: "What I really wanted was an occupation in a field allied to philosophy, . . . one which would provide an opportunity to design and work with automatic apparatus."[22] The typhoid had left his memory impaired, the polio had left him lame, and he seems to have looked to automatic apparatuses as prosthetics. One early contraption allowed him to walk, a "steel brace which I had designed and my brother had constructed at a local blacksmith's shop." Thus equipped, he went on, in the first psychology class he took, to design "a logic machine" of sliding sheet-metal that could solve any syllogism. Unable to continue his studies due to lack of money, he went to teach school in bluegrass Kentucky, where he built a "crude exposure apparatus," a prototype for the Exposure Apparatus he eventually used in his dissertation research. This hand-operated, homegrown device, which he called his "automatic memory machine," was constructed from the discarded scraps of domestic and farm life: tomato cans, rusty bucksaw blades, old spools. It is worth describing in Hull's own words:

> The drum was made from a tomato can fitted with wooden heads. The automatic stepwise movement of the drum was controlled by a long pendulum; the coarse-toothed escarpment wheel controlled by the pendulum was filed from a discarded bucksaw blade. To substitute for a needed gear, a thread was reeled around a large, flanged wooden wheel and then around a spool pinned to the escarpment shaft, and the shaft of the large wheel was turned by a heavy lead weight.

The machine functioned as a primitive image projector, flashing a series of cards at an experimental subject. But perhaps most striking was its aesthetics, reminiscent of Alberto Giacometti's surrealist sculptures of 1925–32, such as *The Cage* and *Hand Caught by the Fingers*. These strange machines are made of crude mechanical detritus that seem to catch or warp inner human urges, so that mental states are rendered as visible outer forms.

Once at Yale, a redoubt of conservatism and privilege, Hull scorned researchers who were outfitted in fancy laboratories yet lacked the know-how to use what was at their fingertips—the remnants of daily life lying discarded around them—and he seems to have held a back-to-the-land view of the laboratory, maintaining that "a person with a little initiative could construct a useful behavior laboratory in a wilderness, given a few simple tools and materials."[23] In 1931 he proposed founding a Museum of Psychic Machines at Yale, but with the institute unsure of its direction, this proposal was never accepted. Hull began to worry that his interest in automata would strike the Yale men who were now his cohort as "a trifle grotesque."[24] Indeed, he worried they would think him insane. So he confined himself to writing about them in his "Idea Books," where he left a lifelong record of plans—eventually seventy-two volumes' worth—for automatic apparatuses built and to be built.

Despite the oddness of Hull's views, the near madness of his system, his theoretical inconsistencies, his untoward use of mathematics, and his bizarre leaps of logic, in 1937 his peers elected him president of the American Psychological Association, from which post he delivered "Mind, Mechanism, and Adaptive Behavior" as his presidential address. On that occasion he unveiled one of his "thinking machines" before a group of psychologists, producing what one eyewitness called an "electric effect" on the audience. But we do not know which machine this was, since there is no mention of it in the published version of the address.[25]

Hull's confidence spilled over to his students, who would be far more rigorous in carrying out experiments than he was. If his system revealed a nexus of swarming activity constituting even the most primitive of reflexes, he admitted to no lurking indeterminacy in this activity, no pause or mystery that could not be dissected and accounted for.

Hull paved the way for a practical science of behavioral engineering. For him, all "higher" activity was not so high after all. His "pure-stimulus acts" were mediated by nothing—no choice, no agency, no thought—and caused an organism to act either one way or another. They "guide and direct the more explicit and instrumental activities of the organism." In short, such acts, "instead of being evoked by ideas, are ideas."[26] Actions are ideas: he did not treat the realm of thought (mind) differently from the realm of action (matter). His point was radical, one that cognitive scientists and artificial intelligence builders are only now, three-quarters of a century later, beginning to understand.

Hull's obsession with building thinking machines from inorganic materials, creating life out of nothing, spurred on a theory that wielded much influence for many years. At the 1951–52 Macy conferences that saw the birth of a renewed human engineering impulse in the social and physical sciences, one participant, the systems logician Frederick Fitch, came equipped with a version of Hull's mechanisms rendered as a logical system. And despite its manifest flaws, Hull's work bore fruit for others, including the philosopher F.S.C. Northrop and the neurophysiologist Warren McCulloch, who went on to play roles, minor and major, respectively, in the cybernetics movement that remade postwar social science. Today his designs are carried out in the language of computer programming.

Hull's dominance over American experimental psychology, mystifying to many, becomes comprehensible when it is considered within the milieu of the institute where it gained prominence. Naysayers like Jean and George Mandler, who deplore Hull's system as a "sorry joke" of a theory, one that was "narrowly concerned with some limited set of response variables operating as a 'function' of an even more limited set of environmental events,"[27] neglect the fact that Hull, as the grandest theorist of this sort, was motivated less by a fascination with "function" than by an exuberance for building new beings from scratch (and by his own leaping ambitions). Working in his laboratory in the wilderness, he paved the way for a science of human engineering, even as he did not carry it out himself.

IT WOULD BE EASY TO CONCLUDE that Hull's system was an absurdity, an extremism born of dogma, a historical accident, and an irrelevancy, since it so obviously reflected his personal obsessions and, behind them, his preference for robotic normality and supercontrol above all. (The Harvard psychologist Jerome Bruner, who as a young scientist witnessed the unveiling of one of Hull's psychic machines, recalls perceiving a "fierce finality" in it that both fascinated and repelled him.[28]) But through the work of Hull, with his most systematic of systems, the institute—first hailed on the front page of *The New York Times* as a great hope for human progress—learned to cultivate its hopes less publicly and advance a hermetic science of the laboratory imagination. It is precisely the strange and momentous singularity of Hull's program that bears further examination. His experiments became part of the "exigencies of reality," in Freud's phrase, because of the way in which they were run.[29] Sometime in 1939 President Raymond Blaine Fosdick of the Rockefeller Foundation arrived to see the gymnastic accomplishments of the institute's rat, but his reaction is not recorded: Did he see what they saw? Was the potential revealed to him, as it was to the scientists, for a new form of authority, a benevolent control that emanated from mechanisms working from within rather than ideologies from without?

Hull's exotic strain of behaviorism came to dominate the institute and spread throughout the nation's social science departments to achieve a level of dominance unparalleled before or since. Which rats would run through which mazes? How would animals become prototypes for the engineering of human behavior? Which pigeons, their eyes sewn shut since birth, would demonstrate "seeking movements" and which would not?

CHAPTER 5

Circle of Fear and Hope

IF YOU GOOGLE THE NAME O. Hobart Mowrer, you will come upon a reasonable number of hits. One dictionary of psychology names him as a pioneer of comparative psychology. A *Who's Who* book lists him as one of history's most influential behaviorists, and another list, posted by the *Review of General Psychology* in 2002, has him ninety-eighth among the Hundred Most Eminent Psychologists of the Twentieth Century (just above Anna Freud but far below Sigmund). Multiple sites credit him as the man who cured enuresis, or bed-wetting, through a behavioral feedback technique of his own invention. There are a smattering of links to alternative-medicine Web pages on bio-feedback. Continue searching, and you will find, buried in the standard descriptions of this now somewhat obscure if evidently respectable personage, a phrase along these lines: "leading innovator in the imposition of avoidance learning conditions." Reading such a dry description today, you could scarcely guess what went on in Mowrer's laboratory and the context—both scientific and social—in which his experiments were carried out. Nor would you be likely to suspect that they have had a tremendous impact on the most intimate spheres of ordinary, everyday life in the dawning twenty-first century. But they have.

Put briefly, Mowrer discovered what constitutes "fear" in the laboratory. This discovery, along with linked discoveries by his cohort at the Yale Institute of Human Relations, formed the basis for the most advanced methods of radically changing behavior in experimental animals and human beings. Even now these methods hold up. In effect, his work displayed how an animal "rewires" itself under conditions of escalating stress. Exactly how coercive could a situation become? To what exact extent and how quickly could one make an animal "learn" to act differently from its ingrained patterns? Mowrer felt that students as well as criminals, normal people as well as sex deviants—for that matter, housewives as well as guinea pigs—could be subjected to his coercive environments and emerge *different* from how they were before.

Mowrer was a key member of a group of young scientists at the institute who experimented in the laboratory to create a new conditioning theory and practice—and who brought it smack dab into the practical matters of everyday life. Their work included developing techniques for gauging and molding human responses: anytime an ad, a message, a political speech, or a sound bite plays on our mounting fears and anxieties, we are in Mowrer's group's territory. Almost all of its members went on to enjoy unprecedented success in their field—six, including Mowrer, became president of the American Psychological Association during the 1940s and 1950s. But during the mid-to-late 1930s they were still working hard in relative obscurity, running experiment after experiment, collectively boosting themselves into a kind of laboratory furor. At first, this regimen was simply in accord with the idiosyncratic demands of their mentor, Clark Hull, who, although hailed as one of the day's great theoreticians of the empirical (and who maintains a decent number twenty-two on the Most Eminent Psychologists list), was not one to carry out many experiments of his own. Busy building psychic machines and dreaming of the ultimate Grand Theory that would outdo Leibniz and other greats who shared his inclinations toward abstraction, Hull was disinclined to undertake the repetitive, day-by-day experiments required to prove his theories, and as far as the laboratory went, "Hull showed an astonishing willingness to delegate work to others."[1] Thus it fell to his students actually to test the Hullian hypotheses. They moved toward experimenting with compulsion, fear, and a scientifically imbued form of despair.

WHEN HE ARRIVED AT YALE in 1934, O. Hobart Mowrer was on his third postdoctoral fellowship. (Then as now, to be on your third postdoc was a dismal predicament, the academic equivalent of being always a bridesmaid, never a bride.) Real professorships were scarce at the height of the Depression, and so Mowrer, having completed graduate work at Johns Hopkins, was forced to take the wandering path of the perennial fellowship recipient. He had in hand a body of work using pigeons to explore "vestibular nystagmus." This was a condition of dizziness induced in a pigeon by strapping the bird to a steel rod and rotating it at varying speeds while hooded or, in some cases, with his eyelids sewn shut from birth (to simulate blindness). Mowrer's thesis work asked how animals orient themselves to space. Specifically, he wondered what factors accounted for their stumblings when they were released from the contraptions. At Yale he spent two more years investigating the "protracted aftereffects of continuous bodily rotation" on dizzy pigeons and then began to broaden his range.[2]

Of course, one could not stitch a human's eyes shut and rotate him or her on a stick. But there were other ways to create an intrusive and dizzying "environment" that would cause subjects to stumble and require them to recover their footing. And this is precisely what Mowrer proceeded to do. His work was so successful, so enthusiastically received, and in such a "hot" area of the social sciences that he presently gained a permanent post at Yale, then at Harvard, and was much sought after for the rest of his career.

Why he was interested in dizziness and the ability of a creature to orient itself in space is another question. Some clues come from Mowrer's own account of how his professional interests grew out of his personal circumstances. In an autobiographical statement, he recalled times in his youth when he felt an unfortunate sense of being "out of touch" both with his own emotions and with other people. At root was a feeling of unreality. All the world seemed a stage, and not a very convincing one, where people played out a false and inscrutable drama; his own part was strange to him, and he felt himself strange to others. He was alienated even from himself. His periodic bouts of such disorientation were so severe that he dreamed of killing himself.

At the age of twenty-three this being-out-of-touch took a more concrete form. Arriving in Baltimore that summer to begin his graduate studies, he found himself "turned around": his mental map of the city could not be made to match its actual geography, and "despite full conscious knowledge that, for example, Charles Street runs North and South, it always seemed to run East and West, and no amount of intellectual effort on my part would alter this stubborn misperception."[3] This hapless condition persisted for the remainder of his time in Baltimore. Accordingly, he began to research spatial orientation and worked, during his first postdoc, with a Professor Fearing, who was conducting research on the effects of "surgical assault" on various portions of a pigeon's inner ear. It was a trying time in Mowrer's life (to say nothing of the pigeons'), but he managed to be competent in one area—the intellectual. Yet even there, as he says, his functioning had a compulsive, driven quality.

Troubled by an intermittent but profound sense of disorientation (emotional, kinesthetic, and sensory), he set out to try to understand these things using his intellectual strengths. He even looked like a classic "egghead"—tall, slim, with an oblong skull and sensitive features. He set out to understand more fundamental problems of orientation, switching from avian to mammalian research subjects. What he made of his research is surprising.

In a key postpigeon experiment at Yale, Mowrer strapped a human subject face-down on a table and hooked his wrists up to electrodes that could deliver an electric shock. (Mowrer does not record how strong the shocks were, but in an earlier version of the experiment, one of the few performed at the institute by Hull, the current ranged in intensity from 3.3 amperes to 7.6 volts, "adjusted to the subject's ability to endure it.")[4] In Mowrer's version, a light that was visible to the subject would flash, and an electric shock would either follow or not follow, at the experimenter's discretion. The initial results found something unanticipated: people started to sweat *before* they received any actual punishment. "It was noted during the early stages of the study that the subjects almost always showed a sizable galvanic response"—that is, increased skin conductivity, a measure of tension— "to the light on its *first* presentation, before the shock had ever been presented."[5] (A significant portion of subjects elected not to continue in the experiment at this point.) The spike in galvanic response happened only when the electrodes were attached; if the electrodes were left unattached—that is, if there was no chance of shock and the subject knew it—there was little or no skin reaction. When the subject anticipated a shock at any moment, anticipation elicited fear as effectively as did the shock itself.

Here was the key: the shock was merely a threat and was not necessarily delivered, or only rarely so. In other words, Mowrer found that the thing the subjects were afraid of harmed them less than their anticipation of it. The pain of electric shock was nothing compared with the pain of expecting it. The point was, not so much that no one had ever noticed this before, but that no one had made a science of it: the dynamic of escalating fear and trembling was observed, measured, and isolated in a lab, and it could be replicated in other situations. The stimulus that the experiment isolated was named the "coercive stimulus" and led to a new kind of response, a state of readiness that Mowrer called the "preparatory set." On one level, this state of dread or tension or anxiety was counterintuitive—for it was the laboratory subject, not the scientist, who, in the most immediate sense, imposed it—but that was also its strength.

The preparatory set caused discomfort ranging from anxiety on up—what Mowrer described as "a mounting feeling of inner tension which [subjects] variously describe as anticipation, dread, apprehension, expectancy, anxiety, etc." That discomfort itself became a conduit for the subject to undergo change. Mowrer discarded the Pavlovian hypothesis that an excited brain state caused an animal or human to re-act to a threat of shock and suggested a much simpler explanation, more in line with what he called general psychological principles: that the spread of stimuli "is a function of the degree of readiness or pre-paredness of a given reaction system and . . . by virtue of the develop-ment of an unusually high condition of readiness, stimuli which would ordinarily be without visible effect are now capable of eliciting the re-sponse for which the pre-existing set was appropriate." That is, it was not simply the strength of a stimulus that determined the effect it would have but the pre-existing condition of the creature itself, its "readiness." The greater the readiness, the more galvanizing the effect of a specific shock or goad, leading to an even higher state of readiness and a growing level of stress. Mowrer did not locate this phenomenon in any particular part of the brain for, like his mentor Hull, he saw it as part of "a given . . . system," a construct that was *physical* but not *physiological*. The subject of the experiment combined with the environ-mental stimuli to form one working system.

IN MOWRER'S EXPERIMENTS, his human subjects reported that their experience of dread would rise to a maximum and then, with the actual occurrence of sharp pain, abruptly subside, producing a kind of pleasure or satisfaction. This appeared to be the case with a bevy of an-imals as well, all of whom could be observed to relax at the point the shock was delivered: "The electric shock seems both in the pig and the sheep to come as a relief, and is followed by a period of comparative relaxation before anticipatory behavior toward the next shock begins," Mowrer wrote. Barnyard animals and "higher" animals alike expe-rienced this cyclical phenomenon marked by ascending levels of nervousness. The spiral would continue to mount when properly managed: "the excitatory value of each successively higher order of conditioned stimulus can be kept great enough"—specifically within a context of dread, excitation, terror, or anticipation—"to be used to establish still higher orders of conditioning." This spiraling effect,

Mowrer suggested, could reach "any desired level." At higher levels, the subject became more and more sensitive to pain; his torment became exquisite. However, should this cycling continue without diminishment, then one would observe, at last, "the ultimate demoralization of behavior": a nervous breakdown. (Leaving aside the question of whether a sheep can be demoralized or suffer a nervous breakdown, it is clear that the experimental subject of whatever species did at some point cease to function in a coherent way.)

In the case of humans, however, Mowrer added an important twist: the whole cycle could be induced with symbols, and such symbolically induced "sets" would follow exactly the same rules as organic ones. Symbols were especially potent if the general level of "motivation" were kept high by the intermittent administration of electric shocks. What this meant was that one could achieve great effects—break someone down or instill a new pattern of behavior—using mostly words, images, and "social incentives," throwing in a bit of unpredictable physical pain here and there.

Mowrer had great ambitions for this work. He felt that the state of unease he had observed in pigs, sheep, rats, apes, cats, and undergraduates—his preparatory set—could become the basis for a unified theory of learning, for it was in this state, he believed, that "true learning" took place most readily. True learning, according to the accepted views of the day (and to a large extent, our own), depended on the need to achieve a satisfying state of affairs. A creature in a distinctly unsatisfying situation would be highly motivated to get out of it, would want to secure relief of whatever kind, and would seek, naturally enough, to alleviate the encroaching agonies of the preparatory set. In short, he would be ready to change himself. Hence an atmosphere of tension and unease (or in its strong form, of terror) was considered ideal for learning: subjects would rather quickly acquire new habits through trial and error and "other goal-seeking behavior."

It may seem odd that the most promising young scientists at this immensely funded institute were busy creating scenarios of "ultimate demoralization," likening them to what could be done in the classroom and on the therapist's couch (and stranger still that Mowrer would later refer to these Yale years as halcyon days). But it is not really odd when one considers the importance of fear and compulsion—that is, highly motivated states rife with unpredictable fears and anxiety-producing signals—in ordinary as well as out-of-the-ordinary life. As we will see, such states are ideal for bringing about greater uniformity of behavior in a given population.

ONE OF THE MAIN PREMISES of behaviorism as it arose in the twentieth century was that an organism and its surrounding environment were really one thing. In fact, this was the premise of the social sciences themselves as they came into their own. The assertion that people are formed, molded, and shaped by their culture or society or upbringing, in various ways, may now sound like a truism, but it was once a radical insight, one that fueled the birth of modern sociology, anthropology, and psychology: the great anthropologist Louis Dumont called it the "sociological apperception" and Walt Whitman was getting at it when he said, "I am a part of everything I have met." (People who understand the sociological apperception are often Whitmanesque in their enthusiasm about declaring its truth; others who have merely had it forced on them in social science textbooks tend to be less excited.)

Particularly in America this view prevailed. Specialists of the new sciences of society did not doubt that the individual and the environment affected each other, but they differed in their opinions about the extent to which this was so. William James, for one, felt that the signal contribution of the Chicago School was to have spelled out this mutual impact of society and the self. Summarizing its insights, James marked out the two key variables that make up all of reality: E (environment) and O (organism). Independent to some extent and interdependent to some extent, these two make up everything that is:

They interact and develop each other without end; for each action of E upon O changes O, whose reaction in turn upon E changes E, so that E's new action upon O gets different, eliciting a new reaction, and so on indefinitely. The situation gets perpetually "reconstructed," . . . and this reconstruction is the process of which all reality consists.[6]

That reality is made, not given, was the great insight of American thinkers in the social sciences like James and Dewey. If you wanted to study reality as a scientist, you accepted this point as true and went on from there.

Experimental-style behaviorists were on a parallel track. They too saw E and O acting on each other without end, but for them these interactions held a special promise: they contained engineering possibilities. To change one meant to change the other. Human engineering could take place with the help of E acting on O and O acting back. Many experiments at the Yale institute focused on James's big and buzzing E (the environmental factors and stimuli surrounding a creature), for this was where any sort of human engineering that would get O to behave in a particular way would start. E was the source and author of O's compulsion. Where the two met and the mechanics of that meeting remained to be seen.

In a 1930 journal article, Hull advanced some unusual ideas bearing on the problem. In describing a typical series of stimuli and responses, he explained how he had assumed the "ever-present internal component" of the subject to be minimal (although its presence was an "undeniable element"). Having thus temporarily put subjectivity aside, he reconstructed what he believed actually went on inside a human being or other organism during any kind of experience. In experiencing a situation, it would seem, the organism makes within itself an exact facsimile of what has happened: "through the operation of a variety of principles and circumstances, the world in a very important sense has stamped the pattern of its action upon a physical object." We are that object. External events of the environment promptly take up residence within us, their receptor:

> Henceforth the organism will carry about continuously a kind of replica of this world segment. In this very intimate and biologically significant sense *the organism may be said to know the "world."* No spiritual or supernatural

forces need to be assumed to understand the acquisition of this knowledge. The process is entirely a naturalistic one throughout. [emphasis added][7]

We carry the world within us even as it acts on us, which accounts for our consciousness of the world—we know it because we *are* it. Hull's view is almost mystical in its postulation of an inner world that is all the more real for being the replica of the outer one. The contents of this inner world, however, are nothing but mimicry. One is reminded of Wallace Stevens's poem "The Snow Man," written around this time in nearby Hartford, in which, hearing "the sound of the land," the listener "listens in the snow, / And, nothing himself, beholds / Nothing that is not there and the nothing that is."

Hull never did get around to restoring subjectivity to his snowman. Indeed, he stripped it away to make room for a miniature diorama of the world. A permeable boundary allowed outside events to seep in and take up residence. At times, said Hull, this inner version may actually supersede the external reality that triggered it. What's inside (the replica) becomes more real—because it is more intimate, because it is known and therefore knowable—than what it mimics outside (the original events comprising the "world").

Hull's view may sound far-fetched, but it predicted by some seventy years current debates in cognitive science concerning how we know the world—whether, in short, "to think is to simulate."[8] We exist as sieves. All experience is an echo, in which the echo becomes more real than the events that triggered it. This makes freedom as it is typically known a slippery thing, for if an organism is merely an ongoing parade of internal reenactments that are somehow more real than their original, how can that creature be said to "act" at all?

Hull's so-called inner parallel, his premise that one can depend on the organism not only to respond to but to live with and in and through its situation, was the seed that Mowrer tended with care. For as Mowrer showed, should that situation happen to be rife with threats or tension, the organism would not only respond to but in some real sense would "become" those threats and tensions. The seed became the tree (and a bitter tree at that). Through his experiments, Mowrer became a transitional figure in the institute's movement toward social control, investigating the ways an environmental "milieu" could transform or destroy whoever or whatever dwelt within it. He could render a creature into just about any state.

LIKE YALE'S LAB RATS, which were being trained to flee the warm places they once preferred to sleep in, the sex deviant could be trained to hate what he once loved. Mowrer's work, as he himself pointed out, could be applied to masochists as well as to schoolchildren (who could benefit from an educational theory and practice that took coercive conditioning into account) and, for that matter, to anyone who experienced frustration or neurosis in daily life. It is hard to change a person's ingrained habits, but this technique held some hope in that direction.

If earlier behaviorists had removed "will" from humans, Mowrer in effect delivered "will" over to the environment itself, via the machinations of the experimenter. Thus Mowrer's "preparatory set" made strides toward a true laboratory model of an environment relentlessly suffusing and flooding an organism. Mowrer created an acute state of tension, a self-imposed punishment that anticipated and in many ways exceeded the actual pain of the worst-case scenario. It was a behavioral augury of the famous Yale physiologist L. J. Henderson's comment, "If people do the right thing, they do not need to be restrained."[9] Mowrer had created a laboratory facsimile of subjection and a prototype of modern brainwashing: a state in which the incentive to avoid anticipated punishment resulted in a willingness and even a desperate eagerness to assume new attitudes in hope of deterring or alleviating the threat.

Modern "scientific" torture—more delicately known of late as "setting optimal conditions," the sine qua non of forceful interrogation—makes ample use of these insights, understanding that the overhanging threat of torture alongside its intermittent imposition and removal is the most effective way to induce a person to cooperate. Likewise, an environment that makes mild and unpredictable "shocks" seem imminent can stimulate a range of subtler changes. There is a word for such an environment, although it was not yet in common use during the 1930s: stress.

More than those of most scientists, perhaps, Mowrer's experiments were a kind of autobiography, for they mirrored the scientist's own state of mind. They were even, at times, therapeutic. In these early years of his career he was, as he recalled, entirely compulsive and driven in his work—in short, stressed out. Indeed, he once remarked that if his life story were to have a title, it would be a fragment from a Rod McKuen poem, "From Torment into Love." In his first few months at Yale he fell into a crippling depression, during which he experienced mysterious and all-pervading feelings of alienation and estrangement. How to give a sense of reality to himself and his surroundings, how to end this suffering, was his difficulty, and it would seem that through the torment of others he assuaged his own. By creating the disorienting environment that his subjects were compelled to inhabit, he managed to orient himself in his own environment. His work was well received and he made a name for himself. Moreover, the atmosphere of collective striving that reigned at Yale, of everyone engaging in a joint endeavor, was inspiring. The fact that, in a seeming contradiction, he remembered his Yale institute days as ones of "bygone splendor and delights" is an index of the degree to which his experiments were transformative.[10]

IF MOWRER REALLY DID FIND RELIEF by imposing his suffering on others (and if he later reversed this process, working to "share" others' suffering and so alleviate it), it is helpful to examine the environmental nexus from which the organism emerged, in other words, to look at his childhood. In 1907 Hobart Mowrer was born into a family of farmers who were just about to give up farming in rural Missouri, outside the county seat, Unionville. His father was forty-three and his mother thirty-nine, considered quite old for childbearing at the time; they had had several children before Hobart, all of whom were grown. It was clear by the time he was six that he was not destined to be a farmer, as he was distinctly disinclined to physical labor and liked books and "talk." In these tendencies his father was not disappointed, and he hoped this last child might become the professional man he had dreamed of becoming himself.

The family moved to Unionville, setting up in town alongside cousins and other relatives in comfortable, rambling houses. Several idyllic years followed for Hobart, and from age six to thirteen he was like a "Little Prince," as he described it, at least in a poetic and psychological sense: "Although my immediate family and I were not a part of what we termed the 'social uppercrust' of our community, there was not a more privileged youngster in town from the standpoint of home, play space, and natural facilities to delight a boy's soul." (These happy days of childhood were never equaled in his later life except by his five years at Yale's Institute of Human Relations, which he described in similarly roseate terms.)

This little world suddenly crumbled when Mowrer was thirteen and his father died, occasioning the figurative death of his mother, who went into a decline that lasted the remaining twenty-seven years of her life. Retreating to live with her daughter in the country, she left Hobart in a shabby boardinghouse to attend high school. This was a "bad 'fall,' " as he put it, for in effect he lost both parents. He became manic and then, after the sale of the family home, fell into a depression when he was a freshman in high school: "Then came a morning when I got up with a mysterious and quite disturbing feeling of unreality and 'strangeness'—about my environment and, worst of all, even about myself." The only experience to which it bore any resemblance was that of two small concussions he had once received, but this time there was no evident physical cause. For two years this went on, he feeling depersonalized, his world unreal. In the last three years of high school, his mother, having returned from the farm, lived with him in modest quarters in town. Emerging from this mysterious ailment, which he would only later learn to call depression, he attended the University of Missouri from 1925 to 1929, where he aspired to "integrate" the two fields of psychology and sociology. (One was outer, the other inner: wishing to put them together was perhaps a way of saying

he wished to integrate his understanding of his life.) Only at Yale, however, would he have a chance to do this, and along the way he brought
his life in line with his work.

DURING THE SECOND HALF of the 1930s, Mowrer and the rest of
the main Yale players based their experiments on Clark Hull's mani-
festos, citing his foundational 1929–34 articles in their own work.
Spreading, filtering, and making its way among the ranks at Yale
and eventually nationwide, Hull's approach acted as a collective primer,
a spur to action, and finally a program so powerful it was peril-
ous to ignore. The group, led by Mowrer and others, even started up
a private club for "Young Turk" experimentalists, membership in
which was to be revoked upon reaching the age of forty (and never to
be issued should the experimentalist happen to be a woman): it
was called the Psychological Round Table and came complete with
mysterious talismans to be carried always upon the person (tokens
marked "PRT" from the Philadelphia Rapid Transit system) and
bonding rituals involving long cocktail hours, the giving of fiercely behaviorist papers, and raucous send-ups of fuddy-duddy senior psychologists.

The key to Mowrer's success was his elaboration of an overarching
state of affairs he called "anxiety," not simple strings of stimulus-
response reactions. He talked about a pervasive effect—an atmosphere, or what he later called a feedback system—that brought
about radical change in the animal subject and did not have to be
strenuously imposed from the outside once it took on a life of its own.
"This capacity to be made uncomfortable by the mere prospect of
traumatic experiences, in advance of their actual occurrence (or recur-
rence), and to be motivated thereby to take realistic precautions
against them, is unquestionably a tremendously important and useful
psychological mechanism," announced Mowrer to his Yale cohort at a
Monday night group meeting in 1939.[11]

Inasmuch as American society has come to live more and more in conditions akin to his preparatory set, Mowrer contributed to the future. With his laboratory apparatus, he built a stressful world that predicted our own: a world in which a fearful shock may happen at any time and in which stress and its effects can actually be engineered, ratcheted up, and in some sense capitalized upon. Mowrer's preparatory set, ranging from anxiety to terror, affects our daily behavior, produces the ideal conditions for encouraging and cementing new patterns of behavior, and along the way reduces that "spontaneous variability of response" that scientists struggled to root out of their rats.

DESPITE HIS SUCCESSES AT YALE and elsewhere, Mowrer fell into periodic sloughs of despond. They could occur at any time, even when things were going his way. One particularly dire episode took place on the eve of his speech assuming the presidency of the American Psychological Association, during which he was so crippled with feelings of "unreality" and the lure of suicide that he had to be hospitalized. Worse, this occurred after he had undergone several hundred hours of Freudian analysis, to no positive effect at all. As he put it, the "alleged advantages of interminable talk" left him as depersonalized as ever and "desperately looking for something to take its place."[12] He was still liable to fall into depression or, as he later saw it, to commit the "sin" of failing to connect with other people. He later felt that not showing up for his inaugural address was a way of forcing himself, against his own will, to "confess" in public to his colleagues that he was an impostor and to expiate his sin of unwillingness to help others. In this and other ways his story diverged from the rest of his Yale cohort, who seemed less than compelled to "confess" to or make restitution for any professional transgressions, and who in any case ceased to be a cohort during the Second World War, coming into their own and dispersing throughout the military and government offices of war, later gaining spots throughout the nation's top universities.

During World War II, Mowrer went to work for the OSS (the precursor of the CIA) and used his behaviorist experience to stress-test men and women for dangerous assignments overseas. During this appointment he reached a turning point in his life. In 1944–45 he attended a seminar led by Harry Stack Sullivan, a former Freudian analyst who had broken with the Freud circle, insisting on seeing people as a function of the relationships they have with other people in their lives. This contact with Sullivan and the Sullivanians caused Mowrer to take the previously unthinkable step of sharing with his wife certain guilty sexual secrets he had carried about with him since adolescence and the onset of his depressions. In the spring of 1945, broaching his sexual secrets had good effects all around: his marriage opened up, his career advanced further, and for the first time in his life he began to feel connected to other people—"significant others," as he liked to describe them—in a profound way. What had been "impregnable," as Mowrer referred to these secrets and, as a result, himself, was now full of possibility.

Mowrer began engaging in group therapy experiments with his wife and some graduate students who had approached him with their problems. (They had come to graduate school, they told him, hoping to alleviate their psychological problems and feelings of alienation, to no avail.) Noting similar dynamics in the difficulties his patients displayed, it occurred to him that a group session might be more effective than one on one. The emphasis was not just on letting secrets out but on "making restitution" for the damage they had caused. It became a movement, called "the new group therapy" or integrity groups, based on the need to be honest with others in one's life and to take responsibility for one's actions and secrets. Unlike the Freudian approach, in which one aired guilty secrets so that they would lose their illusory power, Mowrer felt that guilty secrets were truly about guilt (aka "sin") and that one had not only to confess to them but to alter one's actions to make good on one's crippling debts. Each person was encouraged to understand his or her own personal alienation and the part played by the "pathogenic secret" that prevented real contact with other human beings.

By now Mowrer, who had left Harvard in 1948 to accept a generous research-only position at the University of Illinois, was traveling in different circles: he influenced and was influenced by the Palo Alto school of psychiatry, with its stress on cybernetic communication and how it operates like a feedback system to help or harm people; he made contact with the experimental "community counseling" groups and the pastoral counseling movement; he researched the early Christian sects and tried to recapture the original spirit of Jesus' teachings and those of the apostle Paul, who said we are "all parts of one another"; he came to admire the Freudian apostate Adler for his emphasis on "social interest"; and he rediscovered the writings of George Herbert Mead on the social self. By 1964 he was publishing essays with such titles as "What About Love?" and "The Quest for Community," and by 1967 he was firmly in the emerging group-therapy camp (although the permissiveness unleashed by the *I'm Okay You're Okay* approach was not for him). He was a pioneer in Alcoholics Anonymous, Synanon, and Daytop Village. (The latter two were experimental residential communities for drug addicts.) His integrity groups were not just treatment but a mutual way of life and a form of ambitious social criticism, and Mowrer became something of a crusader. He spoke of a sick society in which people could heal themselves only by admitting their guilt (specifically to those whom their actions had harmed) and undergoing transformation-among-others. If society was making people sick, this was because people believed themselves to be its victim and because they squandered their own will and the goodwill of others, thus making a truth out of something not necessarily true. (Society is not necessarily a total determiner of behavior, but it can become so if you let it.) Unlike any of the other esteemed psychologists invited to participate in the many-volumed *History of Psychology in Autobiography*, Mowrer chose to represent himself with a photographic portrait of his family (including himself, his wife, his children, and his grandchildren) rather than the typical gravitas-laden head shot.

In these groups Mowrer and his wife were engaged in what they considered a "private laboratory," whose atmosphere was changed from Mowrer's earlier laboratories. Behaviorism was transposed to another key: could the science itself be jolted out of its rigid—and in his view incorrect—adherence to "total determinism"? He rejected the spirit of the stimulus-response work that had so absorbed him. It was not that such science was useless or wrong, but simply that it could not yet account for the full depth and particular quality of the human endeavor. (He stopped experimenting with lab animals around this time.) Behaviorism had presented itself as a substitute God, emphasizing the negatives of the Protestant ethic, in which people were weighed down by their sins but were powerless to act to redeem themselves. This doomed free will, he felt, and made each person the victim of a relentless flow of conditioned responses. People were not capable of being responsible even for their own "sins," since all behavior was predetermined by the environmental nexus, and even when they were egged on and egged themselves on to become continually more "free" in their actions, true freedom was elusive in a system of self-predicting inevitability. Mowrer wrote, "Luther gave us *The Bondage of the Will*, which was a frankly speculative and theological work; but now, direct from the scientific laboratories, came a more total bondage: S-R 'bond-age.' "

TO SUM UP MOWRER'S LIFE in a nutshell: he was a man who, tormented by a feeling of unreality, strangeness, and out-of-place-ness, came to find a new approach that was inclusive, critical, and both harsh and kind, and that most of all allowed him to feel at one with others. Around the same time Mowrer was making this about-face, Thomas Merton published *The New Man*, which characterized as a "feeling of unreality" those emotions that accompany spiritual emptiness.[13] Was Mowrer's, then, a spiritual journey? Was he looking in the laboratory first to be God, then later to find Him? The clues he provided afford some cause for speculation.

Experiencing bouts of unreality for most of his life, Mowrer was driven to conduct laboratory experiments on animals to "prove" or understand this feeling. In so doing, he *made it real.* The creatures under his control experienced the "truth" of breakdown, malfunction, crippling anxiety, and debilitating fear—in short, having himself experienced "man's most exquisite personal anguish," as he put it, he spread it around, in an attempt to understand it or to gain a feeling of control over it. In a sense, he imposed his own state on others, and they in turn manifested it. Then, in the middle of his life, he began to find a way to reverse this process and gained control of his life by surrendering it to others. "The study of broken human spirits and the condition of their restoration" now absorbed him, and he was able to enter into human bonds and companionship precisely by sharing the suffering of others. (In his integrity groups, his practice was to have no leader and to show he was no better than others by sharing the details of his own humiliating secrets and subsequent redemption.) This practice seems to have counteracted the earlier experiments, and his new quasi-experiments brought with them a different kind of satisfaction and joy. Once again he communicated his own existential state to his subjects, but in this case (it appears) to good therapeutic effect. His early subjects, by the end of the experiment, had been incapable of anything much at all (having been "broken"); his later subjects, by the end of the experiment, were capable of freeing themselves from the compulsions that had dictated their lives.

He later repudiated the old approach of his Yale cohort, arguing that one could find freedom, not compulsion, in the laboratory if one knew where to look:

> It used to be a great scientific sin to imply that behavior was, in any immediate and direct way, influenced by its effects, its consequences. Now we know that while "a response" is *in progress* information is constantly being sent back to the brain and that what we used to call a response, or *act*, is in reality, composed of innumerable reaction segments which are laced together into purposive action by intricate "control systems," which reflexology categorically excluded. Thus, instead of being merely goaded (the Latin term for stimulated), living organisms become goal-directed, purposive, deliberate, or, if you will, free and responsible.[14]

Even in the laboratory, Mowrer came to feel, self-control could substitute for external coercive control.

The old "push-button psychology" had been about eliminating purpose, deliberation, freedom, and responsibility. It was a form of brain-washing, although Mowrer did not go quite so far as to say so. He did, in another essay, identify a link between Communist Chinese and So-viet "brainwashing" and ecstatic religious conversion experiences (even though he found one diabolical and the other divine). Could this *basic technique*, which he had formerly used to restrict freedom and to con-trol the behavior of laboratory subjects, be used instead to counteract the vast personal alienation many people experience in modern life?

Buried in his late essays is an unasked question: could the experi-mental pigeons and broken-down pigs, the harnessed rats and re-formed masochists, the shuttle boxes and mazes of fear, and finally the social scientists with their countless counts and experimental zeal, could all this, too, like Mowrer, be redeemed?

MOWRER'S OWN PATH led him away from the field of "aversive conditioning." But during the years he spent at Yale, his work provided a strong impetus for merging Freud's ideas into the most advanced stimulus-response theories of the time. This "road not taken," a brief if significant period in Mowrer's life, had large consequences for human behavioral engineering. Soon a final synthesis would come about at the Yale institute through a series of events, coups, and experiments. Then this place, with its absolute adherence to a form of compulsive behaviorism that denied will to everyone and everything within its system, ended up imposing this nullity on the rest of the American population.

CHAPTER 6

In and Out of the South

IN MIDWINTER OF 1947 Simone de Beauvoir engaged in a lecture tour of America and brought existentialism to those who had only read about it. After a few weeks of addressing crowds and attending endless cocktail parties, she took a train alone across the country, marveling at its frigid expanse. In a letter to Jean-Paul Sartre she reported on some reading she was doing:

> I read an excellent book, from which we absolutely must publish huge extracts in *T[emps] M[odernes]*. It's called *Class and Caste in the South*, by an American sociologist—and the method's as interesting as the content. It's a kind of counterpart to your *Portrait of the Anti-Semite* but on the Blacks—and also scholarly in character. It contains everything on the problem of the South.[1]

The American sociologist was John Dollard, and his book had had a tumultuous history even before it kept de Beauvoir reading on the train. Banned in several southern states and in South Africa as incendiary, it was received elsewhere as purveying only the most tepid of views of the color line that separated black from white. Dollard's portrait of a Delta town that he called Southerntown (actually Indianola, Mississippi) was a paradox. With passionate engagement he argued for a more objective view of problems there; with great self-possession he described the fluidity of self and the collectivity of meaning he had found; and finally in a book that resembled nothing so much as a *Gray's Anatomy* of grotesque societal injustice, he maintained that these practices were unlikely to stop, and thus had to be accepted.

At once radical in its insights and strangely do-nothing in its recommendations, the book was as divided as the author himself. A sociologist trained in psychoanalysis and interested in anthropology, he had spent the first three years of his teaching career as the resident iconoclast at Yale's Institute of Human Relations before leaving for Indianola to conduct research for his book. His brand of analysis was unique, made up of psychoanalysis, staunch objective methods, American partisanism, southern particularism, cultural anthropology, literary gifts, and an overarching narrative driven by an adoptive Yankee-Puritan wish-fulfillment. In short, the book was an account of a man who wished to be "stirred," as he believed the African-Americans he met in Southerntown's churches were, to "abandon the structures of the controlling self."[2] It was the chronicle of a man trying to be scientific in a decidedly unscientific situation, trying both to take part and to remain detached. It could even be called existentialist, for its author tried to study social life as a means of manifesting his place in the world.

Following his return north, Dollard underwent a change (both in method and in content, as de Beauvoir might have said). Recanting his earlier work and turning against old leftist allegiances, he took up the pursuit of social engineering and human engineering common to the Yale institute and the Rockefeller Foundation, the two places to which he was most strongly tied. His new goal was to develop a grand theory that would not only "contain everything," as his book on race relations had attempted, but launch a form of pervasive yet subtle social control. Thus Dollard went from being a critic standing on the sidelines of mainstream American social science to a full participant in the project of reengineering human behavior on a mass scale. To "strike a blow for order in the social sciences," as he recalled three decades later, was the aim he came to share.[3] His brand of social science—a way of using "the therapeutic situation" to teach people to live within the limits of their "social maze"—enjoyed a period of tremendous, vaulting success.

This about-face would not have been remarkable and could have been dismissed as a simple case of a talented man kowtowing to the winning views of his profession, were it not for one pertinent fact: he applied his theories and his experiments, first of all and quite deliberately, to himself. He offered himself as a test subject for a thoroughgoing operation in which he started with one set of beliefs and ended with another. It was a tricky business to change your own mind: "The delicate point," he wrote, "is that at which one stops thinking an old line of thought and begins thinking the preferred line of thought."[4] He came to specialize in this delicate point, and while mapping its territory, he became its guide.

IN JUNE 1935 Dollard traveled to the Jim Crow South to write the life histories of African-Americans. In doing so, he escaped the too-close Ivy League environment where he both belonged and did not belong. One of seven offspring of a large Irish-American family, Dollard was gifted not only with a felicitous tongue but also with a capacious mind. Once, he confessed to a friend that he desired to know everything it was possible to know, not for acquisition's sake but for sheer love of knowledge. Studying under a University of Chicago sociologist who insisted that in the interests of objectivity social scientists would "have to spend most of our time doing hard, dull, tedious, and routine tasks" convinced Dollard that this was not a concession he was ready to make.[5] Forthwith he journeyed to Berlin to undergo a training analysis with Hans Sachs, one of the "seven rings" of Freud's circle, at the Berlin Psychoanalytic Institute. On his return, at the age of thirty-one, he had the good fortune to be taken under the wing of the senior anthropologist Edward Sapir, who had recently secured a plum spot at Yale as

Sterling Professor (and who, in thus becoming the first Jewish scholar to gain such an honor, was the first to teach undergraduates). It was rumored that Sapir staked his acceptance of the Sterling Professorship on the administration's hiring Dollard.

The two cotaught an experimental seminar in 1932–33 at Yale that brought scholars from all over the world together to investigate how particular cultures affected individual personalities. This was the now-legendary "Impact of Culture on Personality" project, funded by the Rockefeller Foundation. The group comprised a flamboyant Hungarian, a stoical Japanese, a poetry-writing Frenchman, and a dozen or so others from England, Germany, Italy, Spain, the Balkans, the Near East, India, China, Japan, Russia, Poland, and later Latin America. Each scholar was to be both an object and a subject of the study, since as natives they were products of their cultures' ways and as young scientists they were skilled at describing those ways.

After a year or two of intensive study culminating, it was hoped, in a revelation of how cultures work, the fellows were to return to their "own countries" and "collect . . . data in accordance with the agreed upon plan."[6] But by the end of the first year it became clear that the seminar environment was too superficial and the fellows too inclined toward misty-eyed digressions to really get into the gritty subject matter of how a person becomes who he is: how one is mapped and made, sculpted and shaped by the world into which one is born. (The wayward tendency is suggested by a complaint of the two seminar leaders: "When the discussion turned, as it had once or twice, on psychic research, Halvorsen"—the Finnish fellow—"went mystical and very unobjective.")[7] Thus Dollard's decision to go to the South was a classic on-the-road story, spurred by what he called "academic cabin fever." Bidding Sapir goodbye and bearing mosquito netting that had been a gift from his friend Margaret Mead, he went to the Mississippi Delta for a taste of life in the raw.

HE SET OUT to test the theories he had spent so many years talking about with other academics, arguing, "It is important to be able to say to oneself in a slow and significant way how a culture-personality problem looks when you meet it in nature."[8] "Nature" for Dollard was the town of Indianola. During his five months there, he found the place divided by railroad tracks. On the white side the residents had a pervasive sense of "discipline and order," whereas on the other side the African-American residents "seem to be more on foot, more in motion, and a carefree tone pervades their laughing and joking." It was almost as if they were happier. Not surprisingly, Dollard was drawn to this environment of apparent freedom.

He set out to study only the life histories of African-Americans but found he could not do so. (Going south, especially to the Mississippi Delta, was in vogue for social scientists at the time, but many carried out their studies of African-American culture in isolation from the white culture that bounded and defined it.) The social dynamics between the two halves of town were so fraught and, in the eyes of a northern visitor like Dollard, so obvious and so connected that he could not do otherwise than study them. To try to isolate a "Negro life history" outside the conditions that the town imposed on that life would be ludicrous, for "every fact about Negroes is likely to have an obverse side with meanings for the whites."[9] This simple observation—that you could not look at black in isolation from white—would have many reverberating effects on Dollard's life and work.

The change in his study's aim put him in a dangerous position, however, for a fierce order was in operation in segregated "Southern-town," and Dollard, perhaps with the courage born of innocence, stepped into the center of it. As he wrote Mead from the field, he followed her technique, honed in Samoa and New Guinea, of plunging into the local situation as a participant as well as an observer. But he exercised great caution, too, representing himself carefully to each faction of town. To reassure the white Indianolans, he told them that he was not there to study the economic situation, argue for social equality, or advance propaganda, but

to get some life-history materials from Negroes, explaining that we know very little about how Negro persons mature, in contrast to our knowledge concerning white people. I added that I was acting in the situation as a scientist and not merely as a member of white society, and that my contacts with local Negroes would always be based on my scientific interest.

For white Indianolans, the role of scientist was a guarantee that Dollard would not stir things up. But to black Indianolans, the role of scientist was not assuaging, and Dollard had to present himself otherwise, as their friend.

Forced to use his scientific status as a cloak, to be alternatively removed (for blacks) and wrapped around himself (for whites), Dollard could no longer justify himself *to himself* as a scientist: "I preferred to hold up an image of myself as without affect and objective. This illusion could not be maintained and for the reason . . . alleged, namely, that southerners did not believe it and that it was not true." He had come into a situation so tense with unexamined assumptions and so precariously balanced that he was forced to examine his own motives for being there, to ask, as southern whites did, "What is this Yankee sociologist among all possible Yankee sociologists doing down here studying niggers?":

On several occasions this question was directly and impolitely asked, more often indirectly and courteously, but it seemed to be in everyone's mind. It finally occurred to me to ask myself: What *was* I doing down there? Sectional bias supplied part of the answer. I was there on the old northern errand of showing up the evils of the southern system in its treatment of the Negro, and the suspicion could not be avoided that I wanted to make my research come out that way. The personal aspect of my interest, however, derived from another source, not necessarily a discreditable one, but still a bias. It was what might be called a strong feeling for the underdog, a feeling grounded in my own life history and to some extent previously revealed in self-examination. This resulted recognizably in a tendency to feel with Negroes, to be specially accessible to unusual incidents recording oppressive treatment of them, and to stand with them against the dominance of the white caste.

This "strong feeling for the underdog" had to be put away or at least moderated, for "whatever may be the advantage of such a tendency to the social reformer, it is out of place in the researcher, whose business is to see clearly and report correctly."

This act of stemming personal sympathies in order to arrive at a more "sympathetic truth" (sympathetic, that is, to white southerners, without whose sponsorship he felt he could not have lived there) colored the book, for *Caste and Class* is an intimate account of the process of putting away sympathies while watching them well up elsewhere. It tells the story of a "polite Northerner" visiting a place where he will be forever an outsider, able to live only with an unrelenting "sense of spiritual torsion, willing but unable to conform to the conflicting elements in the social pattern." Dollard's word "torsion," suggesting as it does a body being twisted or wrenched about an axis two ways, gives a sense of his discomfort. The torsion he experienced was certainly between two points of view, but even more so between the pressure to conform, to get along, to make friends with the dominant group, and the pressure to yield to his underdog sympathies, which if acted upon would not allow him to get along at all. These latter sympathies happened to coincide with democratic principles of equality, self-determination, and freedom and furthermore with the larger American legal tradition itself, despite more recent Jim Crow alterations, *Plessy v. Ferguson*, and the repeal of restrictive portions of the Civil Rights Act of 1875, and this made his suppression of them all the more painful. As he said, "The object of study is, of course, precisely the emotional reactions of oneself and one's associates in the concrete social situation."

DOLLARD FOUND THAT THIS SYSTEM of subjugation required active enforcement, which took the day-to-day form of manners and mores. Conformist behavior did not just happen but was orchestrated in small and large ways in order to guarantee the economic, sexual, and social dominance of the white caste.

First there was the brute fact of domination, which was not abstract but personal. As Dollard presented it, the white caste member exacted tributes (in forms of address, in trivial interactions) from those he dominated:

> The tendency among students of culture to consider such acts as tipping the hat . . . or using "Mr." as empty formalisms is rebuked by experience in the South. When we see how severely Negroes may be punished for omitting these signs of deference, we realize that they are anything but petrified customs. . . . In Southerntown the use of "Mr." as a white caste-mark and the omission of it in speaking to Negroes have great emotional value. The Negroes know that to omit the "Mr." in referring to a white man would always mean that the addressee could enforce his right in some uncomfortable way. The main fact is that behind deference from the Negroes is the demand for deference by the whites and the ability to secure it by force if it is not willingly given.

Pinpointing the "sweet submissiveness of others" through which the white caste gained the "feeling of being loved (or appearing to be)," Dollard showed how whites, in the process of extracting such tokens of apparent love—in a continual flow of affirmation from a black man while a white man is talking, or the presumptive use of black mistresses by white men—received them as if they were genuine. Thus "it is an odd thing, but white people seem pretty completely taken in by this behavior of the Negro." For many whites, the appearance became the reality. As a recent scholar, Grace Elizabeth Hale, has observed, in a "dizzying circularity, the mask became 'real' to those who observed it."[10]

Basking in the warmth of the other's regard, the white Indianolan could "forget" that the black Indianolan had little choice but to give it. The result was an odd circumstantial blindness, for even if the white man hoodwinked himself into believing that forcibly exacted tribute was love, the man paying tribute was rarely so taken in. As an informant told Dollard, "the Negro is a 'good psychologist' and . . . he knows his white folks; he adjusts himself to the inevitable and knows how to take some advantage of the situation." Black Indianolans found ways to conform on the outside but not within.

Still, a complication arose: at a certain point the rituals of obeisance required of the undercaste became internalized, further promoting the illusion of having been freely given. In this way a state of mind among black residents was achieved wherein "although deference is demanded, it is also (systematically) given in advance of demanding it." If the white man believed it, the black man might come to feel it— in a move that was both self-protective and self-annihilating. The love that was coerced could become "real" either through repressed antagonism or through identification with the socially powerful white person and the wish to be like him and serve him. As Dollard showed, this phenomenon occurred within an "atmosphere of intimidation" so pervasive and insidious, especially to any African-Americans perceived as a threat (that is, set on educating themselves or their children and exercising their freedoms), that it was almost inescapable; the result was that "every Negro in the South knows that he is under a kind of sentence of death; he does not know when his turn will come, it may never come, but it may also be at any time." In such a stressful situation the only way a person can gain even an illusion of ease is accommodation, especially since he knows that any reprisals will be directed not only against himself but against his family and friends. And this "accommodation" will often, on some level, be "real."

Despite or perhaps because of these insights into the accommodations made by African-Americans, Dollard, having gone South to study the Negro, ended up doing his best work studying middle-class whites. As W.E.B. Du Bois pointed out in a review of *Caste and Class*, Dollard gravitated to the white point of view: "The reader gets the distinct impression that *more or less unconsciously the attitudes of the middle-class white weigh largest in his mind* and that he is not altogether to be absolved from the very usual and widespread suspicion of the testimony of educated Negroes." Du Bois also noted that Dollard gave "the most frank and penetrating analysis of [white] southern mentality which I have ever read."[11] Knowing himself to be at heart a sympathizer with the underdog, Dollard wrote a book that weighed the overdog's sympathies more heavily.

For when it came to analyzing black Indianolans, Dollard had a blind spot. At one point he tells how a black man in a neighboring town, having resisted arrest and shot at a police officer, was chased by a crowd of three hundred with bloodhounds and killed. That night his cadaver was stripped of flesh so that the skeleton could be "arranged and used as an anatomical exhibit" at a nearby (white) college. Re-marking that this last detail of the corpse-as-science-exhibit was "quite horrifying to Negroes," Dollard attributed this feeling not to the specific horror of the spectacle but to a superstitious belief among blacks that "skeletons should be nameless." (He wrote in a letter from the field, "A mob killed a negro about ten miles from here on Saturday, although it was not a scandalous case.")[12] Dollard could or would not admit the particular vehemence of the act: what the display of the man's bones said more clearly than words was that the laws of the land were not impartial, and that only stripped of flesh and reassembled as an anatomical exhibit could such a man, who had been hunted and killed like an animal, become something valuable, a representation of the universal human form.

From the 1890s to the mid-1930s, the act of lynching became more rare but was transformed into public spectacle with wider impact. In this case it became a peculiarly *scientific* spectacle. Despite his hatred of lynching, Dollard would not judge the case in his book: "No one should judge even the most incredible of these acts of violence. We should attempt to identify and understand rather than to deplore them. Unless he's willing to fight no one can judge." If he were willing to "have a personal share" in the struggle to change the system, he would be free to judge and to deplore. This is very frank, and an index of the torsion that gripped him.

AT THE CORE OF DOLLARD'S BOOK is a kind of out-of-body ex-perience. The northern sociologist attended a series of revival meet-ings, and despite the general lack of learning displayed by the local

African-American preachers ("The preacher, of course, does not make such a connected discourse as would be expected by a better educated audience"), he was struck by the collective sentiment there voiced. In a white Protestant church the members tended to sit isolated from one another, but at the revival meetings the participants showed "an obvious eagerness for sympathetic contact, a willingness to be stirred and caught up in a powerful story and to abandon in song, speech and spastic gesture the strictures of the controlling self of everyday life." Dollard confessed to witnessing this group unity and self-abandon "with a certain amount of envy." As it was typical of these meetings to acknowledge any white visitors and invite them to say a few words, Dollard took up the invitation. Here is his description of what happened:

> I determined . . . that if it happened again [i.e., the polite invitation to speak] I would take the pulpit and expose myself to the congregation. It did, of course, happen again and when the next chance came I took it. It was all that I had expected and more too. Not familiar enough with the Bible to choose an opportune text, I talked about my own state, described the country through which I had passed in coming south, spoke of the beauty of their land, and expressed my pleasure at being allowed to participate in their exercises. Helped by appreciative murmurs which began slowly and softly and became louder and fuller as I went on, I felt a great sense of elation, an increased fluency, and a vastly expanded confidence in speaking. There was no doubt that the audience was with me, was determined to aid me in every way. I went on to say that in my country the rain soaked people as it does in Southerntown, that the cold bit through broken shoes in the same way, that poor people have the same desperate struggle for a living, that the scorn of the mighty is as bitter, that the loss of those we love lays a whip across the heart in the same way. This brought a murmuring flow of approbation, of "Well," "Hallelujah," "Isn't that the truth?" and so on. I said then that against these dangers men have always come together to share their experiences and to draw comfort from common warmth and strength. The little talk ended with a round of applause which, of course, was permitted in this case; but more than that, the crowd had enabled me

to talk to them much more sincerely than I thought I knew how to do; the continuous surge of affirmation was a highly elating experience. For once I did not feel that I was merely beating a sodden audience with words or striving for cold intellectual communication. Here the audience was actually ahead of me, it had a preformed affirmation ready for the person with the courage to say the significant word. Of course, there was no shouting, and mine was a miserable performance compared to the many Negro preachers I have seen striding the platform like confident panthers; but it was exactly the intensive collective participation that I had imagined it might be. No less with the speaker than with the audience there is a sense of losing the limitations of self and of unconscious powers rising to meet the unbound, unconscious forces of the group.

To the congregation Dollard could talk "more sincerely than I thought I knew how to" and say that he felt their suffering as his own. He had left the seminar room behind. His moment at the pulpit brought out of him his scarcely realized thoughts. He had found in Southerntown's African-American church a significant whole of which he felt himself a part.

But lacking as he did a place in this whole, as one who both manifested its truth and criticized it, Dollard had to betray the generosity of spirit that had affirmed him and had allowed him to join in. His black friends in Indianola may have realized this even before the book was published. One day, just as Dollard was walking by the post office in the center of town, a black friend marched up to him, as Dollard tells it, all smiles, and extended his hand jovially, as if seized by an overflow of genial neighborly feeling, thus forcing Dollard to take the hand and shake it. Dollard did so, thus committing publicly a clear and, to whites, unforgivable violation of the color line, one that would damage his credibility even among the most tolerant white circles. Even after much reflection, Dollard still could not decide whether his friend had done this in an excess of good spirits or in an effort to "embarrass" him. Today it seems abundantly clear that the man was *showing* Dollard the key to the mystery that confounded him, showing him what it meant, so to speak, for people to "fling taunts in the faces of their executioners." Dollard had written to Margaret Mead about the mystery of how certain Indianolans were able to draw on reserves of courage under extreme duress. ("I haven't got any purchase on it at the moment, except that it seems a very strange sort of action.")[13] His friend asked Dollard to do likewise, to stand up for his sympathies. If he would become involved, he should accept the burden, this man perhaps meant to convey. For the "whole" into which Dollard had been welcomed existed as a by-product of systematic suppression of blacks and denial of their humanity. But the book Dollard went on to write was like his handshake: he went only as far as he was able to.

Dollard saw with terrible clarity the system as it operated on and through caste, yet in the end he upheld it, stating that he would not judge the system even in its most violent transgressions since he was not "willing to have a personal share in any trouble" that "might arise in changing the situation." This unwillingness put Dollard's book in a long line of sociology works that had the effect, as C. Vann Woodward pointed out a generation later, of "encourag[ing] the notion that there was something inevitable and rigidly inflexible about the existing patterns of segregation and race relations in the South; that these patterns had not been and could not be altered by conscious effort; and that it was, indeed, folly to attempt to meddle with them by means of legislation."[14] But despite his disavowals, Dollard did judge the system; for his book, in the very clarity with which it traced the system's lineaments, could not but condemn it. Like his black friends in Indianola, he saw all too clearly, but unlike them, he thought that understanding would be enough. His main advantage, the one that made all the difference, was that he could leave and go back to his other life up north.

He left, spent a year writing up his results, and then released the book. The white Indianolans who had befriended him now spurned him, hating his book, not placated by the "sympathetic truth" displayed therein. This should not have been surprising, for at a certain point late in the argument white liberal southerners would have found themselves becoming distinctly uncomfortable. Dollard as much as says that these nice people may not be sadists, may not go about hunting down others and committing cruel acts, but for all that, they have a part in what goes on. Certainly "there are large numbers of southern white people who do not torture or ridicule," and in fact recoil from those who do, but "their position as caste members . . . is such that they cannot escape a kind of complicity in sadistic actions since such acts serve to consolidate the caste position of the liberals as well as that of the sadistic offenders themselves." Dollard believed that any white person who participated in an eminently undemocratic system and yet called himself a democrat was engaging in a false "disguise" and "extenuat[ion]."

And yet Dollard insisted throughout that the blame fell neither on the dominating nor on the dominated caste but on the system itself. White southern scholars found such claims distinctly unconvincing. As one wrote, "Mr. Dollard's research procedure would seem . . . to satisfy the major requirements of *unscientific* method."[15] Worse than the outrage among southerners was the indifference among northerners. He seems to have believed that his pages would have the force of revelation for his readers; perhaps he hoped they, like the congregation in the Southerntown church, would greet the book with murmured affirmations. When *Caste and Class* was published, however, it garnered subdued responses. In frustration Dollard wrote to Mead,

> There have been one or two rabid reviews by southerners, but in the main the reception of it has been calm. Sometimes I wonder whether I have been deceiving myself and if it really is just old stuff that all the "race relationists" know, as Donald Young seems to think. My considered feeling is that if they know it, it must be unconscious and that the knowledge therefore matters little to science.[16]

Feeling the power of his unmasking, feeling that he had brought to light what had been hidden, Dollard could not understand why others were unmoved and claimed to find its revelations old hat. The reception of *Caste and Class* was a bitter disappointment, although it would become a classic in future years and is still read and taught today.

Caste and Class, as it spins out the twisting strands of Dollard's self-described spiritual torsion, may have a like effect on the reader. The author describes a system for ordering human lives in all its myriad workings, its self-deluding truths, and its clever-to-be-cruel hypocrisy, and so in effect if not fully in word he serves to justify it. This justification, a knot made up of the contradictory strands of the author's personal impulses, takes place at a theoretical and even structural level that is also deeply personal.

Dollard said he was not willing "to take a personal part," but his book amply testifies that he already had, and that is what makes it valuable. Yet Dollard said he could not attempt to judge the "subjugation system" he had studied since it was "likely that such an 'emotional situation' could develop anywhere with the right social ingredients." In this he proved correct, if in a more personal sense than he may have imagined.

IN WRITING *Caste and Class*, Dollard entered the institute fold. A change, really a turnabout, was taking place in him. Previously he had bucked the tendency at the institute to study human relations by running rats through mazes. Insisting that the proper study of human relations should foremost involve humans, he recommended that the institute's scientists study the actual relations between human beings. Holding out for a type of science that would be anchored in the give-and-take of life, he distanced his proposed program from animal studies: "Our projects will deal . . . only with men as 'humans,' " he said rather pointedly. Studies of chimpanzees, guinea pigs, and the neural maturation of children were fine as far as they went but could be connected to a science of human relations "only by very free association."[17] Rats in mazes, in short, had little to do with human dilemmas.

At some point in late 1936, however, just as *Caste and Class* was about to come out, Dollard made an about-face. He would not take over the institute—as he had once fantasized doing, when the institute was adrift and rudderless—but join it. (His letters from this time fairly seethe with hatred for Edward Sapir, his former mentor, who was then dying in Yale's New Haven Hospital.) Repudiating *Caste and Class*, Dollard instead framed the racial situation in the South in laboratory terms.

In an article for *Social Forces* called "Hostility and Fear in Social Life," Dollard portrayed southern society as part of a general social phenomenon that was, at root, more zoolike than humanlike. He depicted the Jim Crow South as a service- and sex-oriented "animal" phenomenon. Those who behaved were gratified and were allowed further exaction of behavior. Each side held its part of the bargain. Dollard now believed the racial caste system was not a matter of injustice or subjugation; rather, racial prejudice was actually a rational response to a limited number of goods for which black and white alike must compete.[18] His earlier work, he decided, had neglected the fact that prejudice makes "sense" in the light of limited resources—someone or some group must be left out. It was not just or unjust—it merely *was*.

If Dollard's flirtation with institute allegiance began as opportunistic, after *Caste and Class* it became a real leap of faith. It was not enough simply to go along with those whom, at least earlier, he had found uncongenial. Rather, he needed to belong. What began as a mask now became a reality. Somehow he elicited from himself a true conversion, so that he was not just going along but getting on board. Exactly how did this process unfold?

THE CHANGE IN DOLLARD could not have taken place without his relationship with one of the "animal experimental people," Neal Miller. Miller's experiments at the institute in the years leading up to the final synthesis were some of the more remarkably creative and boyishly sadistic. A favored Hull student, he displayed an enthusiasm for building practical restraining devices, notably his "special cage" for delivering electric shocks to hungry or thirsty rats and his tiny rat harnesses made out of rubber bands. Late in his life Miller, by that time a distinguished emeritus professor at Yale, devoted himself to defending his animal experiments against those who criticized them as useless, cruel, or both. Struggling to defend a long career of animal research, he listed the concrete gains made by scientists like himself over decades, the most prominent of which included behavior modification of humans to cure scoliosis, bed-wetting, and anorexia, as well as behavior modification of insects to protect crops. He also pointed to a recent program to train pigeons to detect life preservers on the high seas: they detected 85 percent of the colored preservers in one study, he said, whereas helicopter crews only detected 50 percent, and he went on to ask with some rhetorical force, "*If you were floating in a seemingly endless ocean, would you think this research is useless?*"[19]

In 1936 John Dollard was floating in just such waters. In order to "get along" at the institute, he had to change the way he saw things and accept new restraint and authority, just as Miller's rats were harnessed but seemed, after a while, not to mind. As Dollard and Miller were later to argue, one's native "chains of thought" can indeed be broken, so that "necessary chains of thought" can take their place.[20]

The two worked as a team to address a problem besetting the institute group: how to bring Freud into the laboratory. "Neal Miller and I seized hold of a problem which had been suggested in Hull's seminar, namely that the fields of learning and psychotherapy must be connected," Dollard noted.[21] They went about it by asking, "Was there any way in which one could understand the patient's neurotic and maladaptive habits as having to be unlearned and replaced by new and more adaptive habits?" Indeed there was. Although Miller was not brilliant or terribly imaginative—devoted as he was to dogged laboratory work for which, according to one Rockefeller evaluation at least, he showed more enthusiasm "than natural endowment"—he did apparently possess other gifts for certain peculiarly adaptive skills and a type of tunnel vision.[22] These he bequeathed to Dollard.

As the Institute of Human Relations achieved its final synthesis, Dollard was a wholesale participant. Cutting his already-strained ties with old friends from New York like Margaret Mead, Erich Fromm, and Karen Horney, he came out in favor of a social science based on underlying principles of behavior common to rats and men. He freely chose his harness as a life preserver. An internal memorandum reported that he, along with other core group members, now "consistently represented" the institute's point of view in his seminars.[23] Dollard had at last come around.

CHAPTER 7

An Ordinary Evening in New Haven

IN THE FINAL MONTHS OF 1927, the renegade social scientist Harold Lasswell took it upon himself to psychoanalyze a patient while the man was attached to an array of machines. The machines measured galvanic skin response (electrical changes in skin conductivity), blood pressure, pulse rate, breathing, tiny eye movements, and bodily movements; they also recorded the sound of his voice on wax dictaphone cylinders. Bristling with wires and studded with electrodes, the patient reclined on a couch and proceeded to talk about his problems as if he were engaged in a typical therapy session. By employing this setup (aside, perhaps, from lending an added significance to the word *patient*), Lasswell thought he might make history. He would gather information about the inner life that was both objective and subjective. Burrowing into the patient's psyche and contacting the secrets of the unconscious self, he hoped to emerge with quantitative data: charts, equations, numbers, and graphs.

Lasswell's experiment turned out to be a scandal for everyone concerned. Behaviorists hated it because, having labored for years to get hard-to-define things like "ideas" and "unconscious mind" out of the laboratory, they did not want such constructs to creep into their science through the back door. Psychoanalysts hated the experiment because, having worked to penetrate the depths of the human psyche, they did not think the art of its interpretation could be reduced to twitch-recording machines. They did not believe the inner self could or should be graphed. Freudians, especially, thought Lasswell had a lot of nerve hooking his patient up to a machine (one version of which later became the modern polygraph and was also the root of various "scientific" torture devices). When Lasswell's results were published, the New York and Chicago psychoanalytic institutes quickly banned his research.[1]

And yet his crude approach did not die in 1927; rather, it was ahead of its time. In succeeding years, Freud's talking cure and the laboratory's eager machines inched closer together. Could Freud's insights be captured in the laboratory, made "systematic," and redeployed in real life? Could they become an all-purpose matrix for defining and refining the "self"—in other words, a new way to be human—in a modern mass society? They could be, and they were.

Freudian psychoanalysis and American behaviorism came together in a quite specific way, almost ten years after Lasswell's gambit, on a night in the fall of 1936 in a seminar room at Yale University. Startling events occurred on that otherwise ordinary evening, after which five men over the next four years carried out the most intensive and successful work on merging Freud's theories into experimental social science and thence into daily life. At the Institute of Human Relations, where so many laboratory animals had been put through their paces, a true science of behavior now emerged using the tools of psychology and psychoanalysis, as well as biology, ecology, sociology, and anthropology. "The most serious intellectual efforts among psychologists to come to terms with Freudian ideas were made at Yale," the psychoanalyst Marie Jahoda has written.[2] "Coming to terms" meant, in this case, that two opposed ways of looking at the human mind—one that interpreted its profoundest secrets, the other that never needed to interpret anything—began to function as one.

It was more than an arid intellectual encounter between two "schools" of psychology. These two philosophies of the self, in their eventual union, changed how Freudian ideas *worked* in American life. The five men of the Yale group were sociologist John Dollard and experimental psychologists Neal Miller, Hobart Mowrer, Robert Sears, and Leonard Doob, each of whom had also trained in psychoanalysis or been psychoanalyzed. Over the course of four years they deliberately isolated key parts of the "higher mental functions"—mind, self, emotions, psyche, subjectivity—and engineered them in a form susceptible to certain kinds of manipulation. More than that, they spoke to a particularly important audience of midlevel social and human engineering types, the sort of people who eventually bridged academia and more practical concerns. These ad men, workplace counselors, personality testers, pollsters, "applied" social scientists, market researchers, behavior modification clinicians, sex education specialists, and human resources managers were a fairly new class of people, and to them the Yale group offered procedures they could apply in social situations at large.

FOR SEVERAL YEARS the five Yale experimentalists operated in a kind of hothouse obscurity, using laboratory animals to conduct trial after trial and run after run through mazes, down corridors, atop punishment grills, and into other devices that afforded "controlled situations." Dedicated to advancing a grand theory of behavior, they postulated a stripped-down laboratory version of the human being whose interior was free of agency and all ideas (save proto-ideas and the firing of stimulus-response reactions). Experimenting on compulsion, they became very good at creating states of anxiety and fear as triggers to bring about new behavior patterns in different sorts of creatures. But this work was all still in the realm of behavior; it was not clear where the mind of the animal in question, having conveniently been set aside for experimental purposes, actually fit in. Now they needed to explore how to add back an "inner self" of some kind, albeit in a form amenable to their particular science. In this way the laboratory would be more surely connected to the world outside.

Attempts to move in this direction began in the mid-1930s, when the Yale group focused more and more closely on human society, conducting experiments with what they felt were *social*, rather than merely individual, implications. At first this meant simply running two or three rats, rather than just one, through a maze at the same time. For to count as properly social, as their mentor Clark Hull said, a situation must involve "the *mutual* stimulation of at least two organisms."[3] Still, these experiments were not satisfying in accounting for the full subtlety of human social life and man's unique capabilities. However gamely their albino rats navigated laboratory tunnels and T-mazes, scientists still had to admit that there was a "huge leap" between such behavior and complex phenomena like the "bizarre dreams of Freud's patients." But soon the group was able, as they said, to "cast the first rope of a slender bridge across that chasm."[4]

Without repudiating their animal work, they turned to Freud for a way of linking their discoveries with the whole of society as well as aspects of the mind not easy to capture in experiments with rats. Aiming to make a true science for microengineering human emotions, mental states, and behavior, they embarked with a great sense of portent on a series of Wednesday night seminars, with Hull remaining as their leader. They set out to take Freud on. Momentous events ensued—not least, in a rather dramatic coup, the Hullians overthrew Hull, relegating his grandiose obsessions with building psychic automata and quantifying behavioral algorithms to the sidelines and concentrating on their own more pragmatic obsessions with actual experimental control of psychic and behavioral activities. Shifting to a less doggedly theoretical approach, they set out to incorporate the problems of everyday life into their models, hewing to the dynamics worked out in their teacher's theories but sidestepping his idiosyncrasies. Step by step, the researchers made links between machine and mind, rat and human, twitches and dreams. But before we examine how these elements converged on that fateful evening in New Haven, we must trace some of the avenues by which Freudian ideas entered America, an entrance that set the stage for these men's efforts.

IT IS A TRUISM among historians that Americans were pleased to discover everything they were looking for in Freud's theories about the unconscious. According to Freudian psychoanalysis, the unconscious was difficult to reach, a slippery place that eluded full capture, much less full comprehension, but Americans made it easily known. So cheerfully warped from the original was the prevailing American interpretation that the historian John Demos famously called it a downright mistake.[5]

Freud himself was initially hopeful of being properly understood across the Atlantic, and in his first and only visit to this country (to receive an honorary degree from Clark University in 1909) he remarked that America was a place where psychoanalysis at last appeared to be "no longer a product of delusion [but] a valuable part of reality." After the initial flattery wore off, however, he came to deplore Americans' versions of his theories. For example, in 1921, hoping to capitalize on his stateside success, he proposed a series of articles with the title "Don't Use Psychoanalysis in Polemics" for a popular magazine, but when the editors countered with "The Woman's Mental Place Is in the Home," he bowed out of the bargain in disgust.

Freud's own attitude, however, did not deter an eager group of would-be adapters. In the years between the two world wars many American psychologists and neurologists made dogged attempts to translate Freud into something properly scientific—into their own terms, that is. Starting around 1915, a few of the "leading men" (among them the New Englanders James Jackson Putnam, William Alanson White, and Isador Coriat, and the New Yorkers Horace W. Frink, A. A. Brill, and Smith Ely Jelliffe) advanced an optimistic and rather genteel version of Freud. At times they used the (barely digested) vocabulary of Watson and Pavlov to explain Freud's theories, tiptoeing around sex with some delicacy but strongly emphasizing efficiency and scientific management, as was the fashion. Truly, as the historian Nathan Hale observed, "they made the unconscious more agreeable than did Freud," but in so doing these early "translators" distorted Freud almost beyond recognition.[6] They mangled his theories by extracting from his writing certain congenial phrases and formulations that confirmed their own preexisting points of view. By January 1936, when Hull's first Wednesday night seminar convened, the list of American scientists' attempts to "harness" psychoanalysis had grown long.

Meanwhile less starchy Americans expressed a need to investigate the self. The vogue for self-scrutiny first took hold in bohemian Greenwich Village circles, where the talking cure was a source of general fascination at dinner parties and kaffeeklatsches. People there made their own use of Freud. Looking inward and bringing forth for analysis what you found became a serious hobby, a foray into a new area of mystery. Many believed they were pioneers in exploring the realm of the inner self, with its almost endless intricacies, challenges, wayward paths, perversities, and wildernesses. Sometime *salonière* and art patron Mabel Dodge Luhan, for example, held the earliest analytic group meetings in the United States in her dining room. Over the course of her life she produced a many-volumed account of her introspective journeys, characterizing her unpublished "Notes upon Awareness" as "A consecutive summation of the attempts of one manic-depressive character to discover how to free herself of her disability & vacillation, & the various 'Methods' she encountered on her way thro' the Jungle of Life!"[7] By then the terminology of Freud—words like "manic-depressive," "neurotic," "obsessive," "anal-compulsive"—had become second nature among the psychologically up-to-date.

Not only in enclaves of lifestyle radicals but also in mainstream, middlebrow America, people were heeding a call to investigate their conscious and unconscious selves. Cultural historians have noted a widespread shift during these years in how people understood themselves. According to historian Joel Pfister, as early as the 1910s and 1920s average people saw themselves as "psychological," with parts that were hidden and needed to be brought to light. Novels and plays, articles and advertisements, Lillian Gish films and Alfred Stieglitz photographs made it clear that "depth" not only existed but might be plumbed. A process of self-discovery was under way, in the course of which it became evident that the self was located "within" and had certain dynamics and dimensions. With this discovery came the conviction that the self could be changed or controlled: "The production of the notion of an 'inner' self necessitating control coincides—in the United States—with the popularization of the idea of a 'deeper' uncontrolled or unconscious self whose 'discovery' is invested with great meaningfulness and value."[8] It was interesting and, in certain circles, de rigueur to look inward and see what was there. A whole new category of phenomena called out to be cared for, worried over, prodded, soothed, and goaded.

This was the stage, then, upon which the Yale group's drama took place: a popular and professional sense that people had hidden depths to be explored, named, categorized, understood, controlled, and perhaps quantified. Now the work of these five men in four years both sped up and made concrete this process. The quest to discover, know, and control the unconscious was submitted to social science and brought into the laboratory.

THE YALE GROUP was not first in efforts to try to subsume psychoanalysis into behaviorism, but as Lasswell's bumbling 1927 experiment testifies, they were the first to do it well. The story of their success is unusual, for scholars of social and intellectual history usually speak of slow drifts, shifts, discontinuous gaps, and glacial emergences. Yet here was a moment on one particular evening when everything clicked and a new fervor took hold. This moment had consequences for how ordinary life is lived—how Americans think of themselves as particular kinds of "selves," surrounded by and responding to particular kinds of stimuli—that are still played out today.

As months of Wednesday night seminars wore on, experimentalists attempted Hull's "large order" to take on Freud but did not achieve instant success. Big and destabilizing questions on the neobehaviorist side kept any bridge to Freud from being built. Up to this point Mowrer's, Miller's, and Hull's work had described the laboratory "environment" and how it could exert a molding or coercive influence on individuals. A society or civilization was in some sense a much larger environment that could exert similar stimuli or cues on those who shared its domain. But what was the nature of the stimuli? Were they palpable? What form did the response take? And most urgently, *why was the response not more uniform?* Hull had pointed out that "the relative similarity in the stimulating situations encountered in the lives" of organisms was "usually considerable."[9] If this was so, if even personal experiences were not terribly varied—for Hull believed he had shown that the inner life was in fact a microcosm of the outer play of stimuli—why were actual people's responses sometimes so much at odds, sharing as they did the same "stimulating situations"? Why will one bystander dive in the river to save a drowning cat and another stay ashore? Why does one brother become an outspoken anarchist-activist and another a close-lipped investment banker? Hull's reliance on the stern logic of "hypothetico-deductivism" made it difficult for him or his students to grapple with such problems.

Hull was, let us say, uncomfortable with the messy human problems of day-to-day life, and so his seminars in the early months of 1936, which had witnessed the first "systematic" attempts to incorporate Freud into neobehaviorism, ran aground. There was some muttering in the ranks. As one of Hull's students recalled, "A new approach was needed."[10] Unwilling to abridge the system he held dear, Hull admitted to having doubts about the entire project, preferring to abdicate anything that threatened his own large theories and asking with a note of hope, "On the assumption that psychoanalysis leaves much to be desired in the matter of theoretical structure, does it follow necessarily that this is a permanent defect?" Hull believed that if Freudian ideas did not yield a "workable postulate system" and result in lovely "derived coherent theorem sequences," the group might just as well look elsewhere.[11] No logarithms, no Freud. Beginning to suspect the near-fanatical depths of Hull's commitment to Hullianism, his students were of a different opinion.

Late in the fall of 1936, with the light fading outside the institute's windows, the closed circuit of Hull's system was forced open—or one could say, the younger brood, finding the more direct path A blocked, at last took path B. At one of the Wednesday night seminars, Hull's student Neal Miller and John Dollard together advanced a hypothesis that had the force of a bombshell. Instead of importing Freud wholesale, they explained, they would proceed concept by concept, which would allow them to incorporate the Freudian instinct or *Trieb* into their own model. They debuted the so-called frustration-aggression hypothesis (also known, in time, as the Dollard-Miller hypothesis), which sounded quite simple at first: when people experience frustration from one source, they lash out elsewhere in the form of aggression. Examples of aggressive events that took advantage of preexisting frustration were football games, lynchings, strikes, wife-beatings, sibling jealousies, the reading of detective novels, and war. Now if you took frustration as the *stimulus* and aggression as the *response*, this very common phenomenon was suddenly capable of being expressed in purely behaviorist terms. You could even show it happening in laboratory rats: place them on an electrical grid so that they are frustrated in their attempts to attack the source of their pain, and they will "lash out" with aggression, first at each other and later at any convenient object, including a human celluloid doll provided for the purpose. In another experiment, white American boys at camp, placed in a frustrating situation, were observed expressing increased hostility toward Mexicans and Japanese.

Most of those present for the unveiling of the Dollard-Miller hypothesis characterize it as a bolt of lightning, an exploding bomb, or a conflagration. So striking was it that, as one participant recalled, "Nearly everyone caught fire with the frustration-aggression idea." Lives changed course that night, and the seeds were sown for some careers to be made and others destroyed. From that night on and for the next four years, "the Yale Institute of Human Relations was . . . probably the most exciting, stimulating, and productive enterprise of its kind in the world."[12] Within a few years the hypothesis was functioning as a lingua franca among psychologists and other social scientists, and it achieved a level of acceptance unmatched, perhaps, in the history of the social sciences: it was accepted by just about everyone in the field. During the 1940s it was considered a tested triumph of American psychology, its thesis used during World War II in campaigns to help motivate troops (by directing already-existing frustrations toward proper objects of aggression) and afterward in civil affairs to instruct reconstruction officials in Japan (refocusing racist aggression away from the Japanese).[13] In the 1950s, as we will see, its uses multiplied still further.

Still, more recent investigators have found the impact of the frustration-aggression hypothesis to be hard to fathom. The papers of these key meetings have never been published, but two scholars who recently scrutinized them found the discussion "arbitrary and . . . bewildering," alternating between high-theoretical jargon and everyday examples given the weight of profound insights. Coming across the hypothesis today engenders a response closer to "duh" than "aha!" Content-wise the new hypothesis was not really new, even then (as its authors themselves admitted); more to the point (as its critics contend), it was not really Freud and was "in essence inconsistent with Freudian theories."[14]

But it did do one thing: quite simply, it allowed behaviorists to bring the deep inner recesses of the mind back into their theories. The inner self was now seen in behavioral terms as any number of processes, interconnections, and physico-chemical reactions. The new hypothesis allowed them to continue their rat-in-maze research and their emphasis on social-science-as-pure-science even while engrossing themselves in the problems of passion and the darkest human desires. They now viewed these problems and desires as dynamic and economical, which meant they functioned by rules and flows and could be fully explained—and soon, perhaps, even measured in hard numbers. So it was that the frustration-aggression hypothesis acted as a wedge to open up the field of Freud. Earlier approaches had been too timid, relying on integration, translation, or verification. Now it was absorption, whole hog.[15]

THE YALE INSTITUTE made a stripped-down and easy-to-use version of Freud. Indeed, the 1939 volume that resulted from its members' collective efforts, *Frustration and Aggression,* was remarkably down-to-earth, using such humble examples as Johnny choosing mince over raisin pie to describe the absolute necessity for the individual to adjust to society's demands and like it. At heart the hypothesis was an imprimatur to talk about everyday life, to bring theories close to people's lives. It was social engineering writ small (geared to a human and unprepossessing scale) and human engineering writ large (for each human act was to advance the larger cause of collective social order). Its folksy tone aside, the work was astoundingly ambitious, offering nothing less than a new way to enforce social control within a large-scale society, to bring a greater degree of conformity to the system, and to assist people in choosing to have no choice in the matter. In this way the Yale group carved out a "drastic extension of the therapeutic terrain," as one historian has noted—drastically extended, that is, from the laboratory maze to the social maze.[16]

Now representatives of the fields of sociology, social psychology, political science, economics, and anthropology entered into and merged their concerns with those of the experimentalists. Anthropologist George P. Murdock's universal files distilling and cross-referencing all facts of all known cultures (or a representative sample thereof), known around the institute simply as "the files," became an integral part of the institute's program, and promised a wider range of human norms and cultural forms to study than the animal research alone had. Psychologists and sociologists trained in psychoanalysis, such as Dollard and Earl Zinn, had new power, while most of Hull's students went and had themselves analyzed in New York or Berlin. Studies of "conflict," "frustration" and "aggression," and "anticipation of punishment" proliferated in these fields. The institute's views of why people do what they do continued to crop up, for example, in policy documents and administration manuals. During World War II, when anthropologists and sociologists evaluated the situation of Japanese-American evacuees at the Poston Relocation Center in Arizona, they used the frustration-aggression hypothesis as the administrator's tool par excellence for analyzing tensions in an interned and unhappy population and for determining how to dispel them. In succeeding years, many stripes of social scientists, from counterinsurgency experts to on-the-job counselors, came to speak of aggression as a hydraulic phenomenon that welled up and could be dammed or diverted—never mind its specific causes—which contributed to a view of the world as a place to be modified and engineered rather than understood and entered into.

A tricky climate prevailed in the case of a penurious but talented young Freudian named Erik Homberger (part Jewish, part Danish, and later better known as Erik Erikson) who in 1937 accepted an invitation to the deep-pocketed institute as someone who had learned his Freud in Vienna "at the source." Homberger offered a direct link to the master, a brain to be picked and energies to be tapped for the cause. Once he was in New Haven, Homberger felt tremendous pressure to hew to the preferred line of thought and to add his efforts to the common task of translating Freud into behaviorist terms. He found this work unpleasant, and in particular he objected to what he felt was the precipitous co-opting of Freud's subtle theories and the equally precipitous publicity tactics—for Hull was sending abstracts of the procedures to psychological departments throughout the country. After a year, Homberger quit for California as institute members made catty remarks about the Dane's daintiness in his wake. Despite Homberger's dissatisfaction, the new program held complete sway at the institute and gained a foothold in social science departments across the country.

This coalescence made up the core of the institute until the eve of Pearl Harbor, after which its members dispersed to war service. They had finally moved from Hull's broad principles to the microdynamics of the inner life (which in turn ran on the "effects of anticipated punishment for aggression") and achieved a synthesis that changed the course of American social science for the next thirty years. The work of

Mowrer, Miller, Dollard, Sears, and Doob made Freud's ideas approachable and easy to use for the mainstream of American experimentalists. However, this advantage had its downside:

> Freudian terms and crudely analogous observations invaded the experimental literature on a scale never before attained, but the price paid was that Freud's concepts were turned into vague conceptions, barely related, and at times actually contradictory, to their original forms.[17]

Like their American forebears, the Yale five distorted the distinctive qualities of Freud's work. And yet it was not explanations of Freud they were after—they wanted equations.

What types of equations exactly? As we've seen, they believed that the behavior of animals in laboratory mazes and cages was *equal* to that of humans in the world, but this belief required proof. The "analytic situation" of therapy was a "learning situation" akin to those found in the laboratory, which established a valuable parallel: just as white rats *learned* in laboratory mazes to adjust themselves to the demands of their environment and its "realities," so too might people. Scientists believed that people who did not cooperate and conform to prevailing norms were simply afraid to do so. The solution was to make them obey authority not by force but by desire, through a type of conversion configured as an environmental stimulus-response reaction. The "analytic situation" was one of the best places for this to happen. There the patient learned to accept the analyst's view of reality as his own and, in coming to see the world as the authority figure did, concluded he had made his choice freely and had not been coerced.

If the analytic situation was a miniature laboratory for engineering one person's behavioral responses, American culture itself was potentially a larger laboratory for engineering people en masse. Many experiments with albino rats, babies, Boy Scouts, and adults followed, both in the laboratory and in analytic situations. Everything became a matter of cues and miscues, imitation and nonimitation, learning and punishment. The Yale group devised a four-step process, deduced from animal research, that they claimed drove all humans in their pursuit of an adequate life: "the individual must want something (drive), notice something (cue), do something (response), and get something (reward)." That was all. Drive-cue-response-reward: rats proved it in the laboratory maze and humans acted it out in their own cultural mazes.[18] Drive-cue-response-reward was the endless and inescapable staccato of life.

ACCORDING TO THE AUTHORS of *Frustration and Aggression*, life was, basically, a running series of frustrations. Birth ("possibly one of the earliest frustrations") is followed by a period of infancy, in which all sorts of sought-after gratifications are interfered with. Down the road, childhood comprises such a gamut of frustrations—children must be trained to be clean and tidy, to be less clingy, to avoid masturbation, to be gender appropriate, age graded, and properly school-bound—that it could almost be defined as an inherent frustrator. (Childhood, as the authors say, is "a period of persistent, forced, and sometimes violent, changes in habit.") Adolescence ratchets up the frustration level, as seen in increased aggression and irritability. Adulthood, while easier, "is not lacking for frustration" either, as one encounters marital demands for adjustment as well as career and social mobility requirements. Death, when it comes, marks the end of frustration, but waiting for death can itself be a further source. ("Death is the final interruption of all responses and its anticipation is often attended by a feeling of futility and the sense of a life unlived.") The beginning and the end, like everything in between, are unsatisfying, and an individual constantly channels these goals into different activities— "satisfying goal-responses"—in an attempt to reduce frustration and achieve a modicum of satisfaction.[19]

The authors suggested that some lives are less frustrating and therefore more satisfying than others, and they painted a sort of Caliban-meets-Rockefeller picture of social options. For example, a captain of industry is said to be the person least likely to commit a crime, for he has achieved the ultimate "goal-response" of success and is able to meet his basic needs for food, sex, and other essentials. To have some stake in prosperity and something to lose diminishes the criminal urge and even frustration itself because, logically, an expression of it imperils that hard-won success. A man in his mid-twenties would therefore tend to be less aggressive or frustrated since at this age he would be capable of worldly success. A woman, on the other hand, would be most likely to achieve her version of success at a younger age by marrying an older, successful man; she thereby gains "impulse gratification" before her male contemporary. But in a society where her status depends on physical attractiveness, her tendency toward criminality may be expected to rise as she ages (raising for this reader the specter of a phalanx of criminal Blanche DuBoises).

For some—for example, those with gross physical deformities—fitting in is quite evidently not an option. A villainous appearance might produce villainous behavior—not due to old-fashioned Lombroso-esque attributions of inherent evil, the authors hastened to add, but because "offensive-looking individuals" will tend to encounter frustration "in attempting to pursue many of the socially desirable walks of life" and thus will be "prone to gravitate to the 'underworld.' " In a self-fulfilling physiognomic prophecy, normality (looking normal) produces normality (acting normal). Unfortunately, the abnormal or the deviant sends out the wrong "cues" or "stimuli," and for them normality therefore is out of bounds. Those unable to fit in tend to become deviants who must be watched.

In the Yale group's locutions, frustration became a free-floating motivator for crime that automatically settled on those unable to fulfill the goal of fitting in, and any type of "marked aberration" counted as a form of frustration. Thus "it is not surprising that divorcees—who are often more or less chronically maladjusted in other ways—should show an unusually high crime rate" due to their necessarily high level of frustration, more or less sexual in nature. (The authors speak of the "exceptional frustration" caused by the loss of sexual fulfillment from one's husband.)

Women who are unmarried after thirty-five, divorcees, unskilled workers, the foreign-born, bastards, American Indians and Negroes, those with unwholesome home conditions, those living in "demoralized neighborhoods," and city dwellers, along with all the physically deformed and personally unsavory, will accrue this frustration factor disproportionately (as statistics of crime seemed to confirm). Outsiders and the unassimilable are potential threats to society, damned by definition. They are frustrated, and frustration must, of its nature, lead to further action of expression or substitution.

Thus, according to *Frustration and Aggression*, the purpose of life is to achieve a normal life. (Normal, bourgeois family life, however, and the demands of civilization itself, were precisely what Freud in *Civilization and Its Discontents* found to cause painful conflicts and suffering capable of rising, on occasion, to the level of tragedy.) Here, instead of a tragic arc, we have a constant calculation that goes like this: Frustration accrues to everyone, but those whose "performance of goal-responses" places them safely within the normal will attract less of it, will stand out less, and will slide more easily through life, thus benefiting themselves and everyone else. Ultimately, in the eyes of the institute workers, frustration was the blocked desire to fit seamlessly into one's place, and they speak unwincingly of "pegs" needing to fit into their proper "holes."

The argument presented itself in a guise of evenhanded unobjectionability and in fact was the product of good-hearted liberal progressives backed by hardheaded laboratory science, but it became a brief for the suppression of any evidence of life itself. All activity in the world boiled down to optimizing one's stimulus-response reactions. No matter how deep inside you looked, it was stimulus-response all the way down. Since the "goal-response" is a given (to run the social maze and so to feed the smooth running of the social machine), to fail to meet that goal-response is to be left by the wayside, to become immediately and by definition suspect, a statistical risk for criminality and—since correlationally this outsider status is as often as not a result of deviance or freakishness—therefore an enemy or potential enemy of right-thinking Americans. It was a social science of dull despair, clothed in garments of optimism.

THE YALE SYNTHESIS of Freud and behaviorism gives insight into a peculiarly intimate view of the normalization process afoot in America. The same year *Frustration and Aggression* was published, the following item appeared in *Time* under the heading "People":

> Because he was tired of having people doubt that his name was really *Yale Harvard Pinsker*, Bronx salesman Pinsker legally changed his name to Yale Harvard Perry.[20]

Pinsker renamed himself Perry to suggest Ivy League associations. Are you sending out the wrong "cues" for your desired response? Change your cues. The subjective element of self added to this science was a matter of a fine attunement to the expectations of others. Such operations embedded themselves in the institute's laboratories. The world, like the laboratory, was made up of stimulus-response reactions, and the stimuli of these ubiquitous reactions were physiognomic cues. In order to fit in properly, you must work to diminish the response to your black skin or Jewish nose and downplay them, giving off less in the way of stimuli. Personal identity was a collection of cues, which could be tuned to stimulate a new response. All could be Perry, not Pinsker, or at least try.

These stimulus-response therapeutics allowed the Yale group to strip away things like Freud's "urge to freedom," and the particular situations of working-class, immigrant, or downtrodden members of society, and even "life" itself. These conditions figured as sand in the gears just as much as any abnormality or grotesquerie. Even as the Yale group gave society priority over any residue of individual freedom, they phrased the argument in terms of individual fulfillment via goal-response—that is, freedom to fulfill oneself in acceptable ways. One must learn to want within the limits provided, but compulsive wanting is certainly to be encouraged. "The aim," as James Baldwin once put it, seeming to channel the Yale view, "has now become to reduce all Americans to the compulsive, bloodless dimensions of a guy named Joe."[21]

A key element of this reduction was to encourage "substitute responses" to the dilemmas posed by civilized life. Since the white picket
fence could not contain everyone, and those left out stewed threaten-
ingly on the margins of America, the logical solution was to cultivate
alternatives—substitutes—for those who couldn't have the real thing.
Already "substitute responses" were quite prevalent, and the Yale cadre
believed that actual Freudian-behaviorist mechanisms could be isolated under laboratory scrutiny. An attendant experiment by two of
the group's members, Doob and Sears, extended the *Frustration and
Aggression* argument. Administering a questionnaire to Yale undergraduates detailing fifty-three situations deemed annoying or frustrating, they made a typology of the reported substitute responses. A
"pure" substitute was one in which a man loses his job and, instead of
lashing out at his supervisor or going on strike, "indulges himself in
some small luxury" like a cigarette or a pastry. Another substitute response was the "nonovert" or "conceptual" expression of aggression.
("I'll kill him in my dreams.") Ultimately Doob and Sears hoped to
forge a quantitative theory for these different substitutions, but they
felt ready at the time only to point out certain factors determining responses, the central one being "anticipatory responses to punishment-
for-being-aggressive." (The only other factor they mentioned was the
strength of the frustration itself.)[22] Substitution was a likely approach
to controlling social life, and since anticipation of punishment was the
main factor determining substitution, the authors concluded that so-
cialization as a process *consists of* such anticipation of punishment: "If,
even though individuals find a mode of behavior very satisfying its occurrence is simultaneously inhibited, then it is reasonable to conclude
that *they have been socialized to the extent that they are able to anticipate punishment from acts which they also anticipate to be satisfying.*"
The social man is the scared man, the man who learns to satisfy himself by substituting small rewards—a piece of cake, a cigarette, some
petty hatreds—for a great and unrelenting fear. In a 1941 presentation
to the group on "The Freudian Theories of Anxiety: A Reconciliation,"
Mowrer discussed anxiety and pain reaction and showed how, when a

low-level punishment is actually present all the time, anxiety itself becomes a learned response.[23]

THE WORK OF THE YALE GROUP finally addressed the heretofore-denigrated realm of human subjectivity. It cemented the conditioning of animals to the human condition and reintroduced a certain circumscribed simulacrum of freedom or choice into the laboratory model, its conceptual stimulus-response framework having initially excluded *any* freedom from compulsion. (Similarly, through the element of "choice" in consumer capitalism, people are encouraged to believe that their selection of a particular product from an array of products is not only an act of self-expression but a subversive flinging aside of the very bonds that compulsive consumption reinforces, as in the maxims "Be different," "Think different," "You're unique," "Indulge yourself," and "You're an individual just like everyone else.")

As with certain pills, the effects of the Yale model were both immediate and sustained-release. Broadcasting the program far and wide, the Yale group looked on with satisfaction as many other researchers adopted their approach. By the mid-1940s and 1950s much research along the lines of their program was under way, producing more than four hundred laboratory studies that rendered Freudian phenomena in behaviorist terms, with at least a thousand more published by the 1970s.[24] During the 1950s, 1960s, and 1970s, a mechanical and distinctly un-European Freud invaded the general literature, the inner sanctum of American psychoanalysis, and popular culture itself.

One long-term and far-ranging result was a new way to organize the individual human organism in relation to its environment, in what theorist Nikolas Rose called the "governed soul" and historian T. J. Jackson Lears the "managed self," with its widespread promotion of managerial values "systematiz[ing] an anxious, driven mode of personal conduct."[25] The self is "governed" and "managed" by an environment that continually molds and shapes one's behavior, thoughts, dreams, and innermost longings. These longings are, quite often, the inevitable responses to being nudged, tuned, and aligned in one direction or another. Freud may have died a scientist and doctor (a death marked recently in *The New York Times Magazine* by the announcement of his rebirth as "The Literary Freud"), but in "behavioralized" form and stripped of any remnant of Viennese elegance he still inhabits our every impulse. It would not be far-fetched to describe our living, working, consuming, and recreational worlds as an elaborate Freudian Habitrail.

It was in America, as nowhere else, that human engineering merged with psychotherapy.

PART THREE

Files:
Out of the
Laboratory

CHAPTER 8

The Biggest File

IN 1951 SOCIOLOGIST C. WRIGHT MILLS described the anxieties of the modern age by invoking the image of society as an enormous file. He was perhaps unaware of the existence at Yale University of an actual filing cabinet of just such proportions. Built in 1928, named the Cross-Cultural Survey in 1937, and renamed the Human Relations Area Files in 1948, these files were meant to classify and then catalog all knowledge about humankind. With them, their founders intended "to permit the ordering of information on man's various environments—including climate, geography and topography, flora and fauna, as well as the physical, social, and behavioral characteristics of a people, their beliefs, value systems, religion, and philosophy." With scientific steadiness and bureaucratic efficiency, they were to amass this range of data not only for the white-collar world but for all worlds, beginning with "a representative ten percent sample of all the cultures known to history, sociology, and ethnography" and ending, it was hoped, with a significant proportion of the world's information filed in a single big box.[1]

In examining the origins of this project almost three-quarters of a century later, it is difficult not to be amazed at its inaugural hubris. The anthropologists and sociologists involved appear to have set out to provide total access at any moment to a universe of facts. Furthermore, in the name of efficiency, they sometimes sounded as if they were hawking a new kitchen appliance: the files, as their creator, George Peter Murdock, characterized them, were a terribly useful time-saving device for the scientist or scholar who hitherto had been forced to "ransack" the literature but who could now "secure his information in a mere fraction of the time required to [consult] the sources for himself." Having at his fingertips "a representative sample of cultural materials . . . for ready accessibility on any subject," he would be able to use the files to answer social or cultural questions "not one by one but in quantity." Indeed, old-style research in books might soon be obsolete, for "the object has been to record the data so completely that, save in rare instances, it will be entirely unnecessary for a researcher using the files to consult the original sources."[2] The sum of the world's contents was to be converted to text-based code, stored on file cards, and maintained in a systematic fashion so that the resultant data could be recombined or extracted or processed at will—and all this without a computer, much less a World Wide Web.

If the enormous file, the incorporated brain, and the great salesroom were Mills's images for grasping the twentieth century's "new universe of management and manipulation," Yale's file grasped the world more directly. It held in its confines the essential data of many cultures—40 in 1938, 150 in 1949, 285 in 1966—with the goal of making these data amenable to statistical and correlational methods of scientific inquiry. By midcentury a large wooden double-tray Remington Rand "Aristocrat" file cabinet, Grade A, held several million five-by-eight-inch cards, each one carrying a bit of culture, each bit coded for easy reference and cross-indexed to other bits of culture. By the mid-1960s the number of cards it housed had grown to 65.8 million. Unlike previous anthropological endeavors, this one did not just depict a single specific culture among the great array of cultures; *it was itself a picture of the array.* In time, as the cold war set in, this picture was used to give precise dimensions for conformity of behavior in public and personal life—it showed where modern practices stood in relation to how other people in other places lived, and it gave a scientific basis to what was considered "normal." As such, it illustrated rather graphically the steps by which Americans would come to know the world and themselves, and the two in relation to each other.

"YALE'S BANK OF KNOWLEDGE," as their promoters sometimes called the files, was at once infinitely expandable (in practice) and entirely complete (in conception). In some ways it was nothing new, for in the nineteenth century scientists quite commonly dreamed of making a true science of cataloging and comparing cultures. In 1898 a Dutch anthropologist, S. R. Steinmetz, set out to file the vital elements of fifteen hundred cultures but lacked the technological means and clerical assistance, having only one or two "lady helpers." Anthropologist E. B. Tylor made a similar attempt and classified all existing marriage and descent institutions in his 1889 article "On a Method of Investigating the Development of Institutions," but kinship was after all only one of many parts making up a culture. Herbert Spencer in his *Descriptive Sociology*, and later William Graham Sumner and his students in *The Science of Society*, tried to draw up a full outline of the totality of culture, but neither man had been able or inclined to file every bit of cultural data underneath it. Most of these efforts failed due to a lack of secretarial zeal and adequate filing techniques.

Their common dream of a "roomful of drawers of notes" filing "all data on all peoples of the world" was finally realized at Yale—with, however, a difference.[3] Almost from the outset, the Yale files were a team project, big in scale and about to be bigger, filed and cross-coded in new and systematic ways. This meant that they were easily transformed by the impetus of World War II into an intelligence source and collator, an administrative device for tracking displaced or interned people, an aid to military occupation, and eventually a cold war strategy tool to use in locations at home and far from home. A leapfrogging of acronyms marked this transformation, from the files' foundation at

Yale's Institute of Human Relations (IHR), their formal organization through the *Outline of Cultural Materials* (OCM), their physical realization in the Cross-Cultural Survey (CCS), their wartime utilization as the Strategic Index of Latin America (SILA), their postwar coalescence as the navy's Coordinated Investigation of Micronesian Anthropology (CIMA), and their ultimate systematization as the Human Relations Area Files (HRAF), all capped in 1953 by an "Inc." No scholar could have foreseen such a lively future for what was, at heart, a big encyclopedic device.

The files' creator, George Peter Murdock, was himself a transitional figure and embodied a mix of old and new American styles. Although raised a blueblood Connecticut Yankee (a great-great-great-great-grandfather, Peter Murdock, immigrated around 1690 from Scotland and settled on Long Island, with his son John and all other ancestors moving to Connecticut), he went on, in the course of his life, to follow the sociological formula for the white-collar worker and was "always somebody's man, the corporation's, the government's, the army's."[4] Murdock worked for all three, in addition to the university. His life began in the patrician and agrarian setting of his father's farm and seems to have ended in a library: as a professor he was renowned for his encyclopedic knowledge, gained by spending nights from 5:00 p.m. to 8:00 a.m. in the stacks, in addition to performing his regular teaching duties. An obituary by one of his colleagues noted, "In the course of his research, Murdock acquired a more exhaustive knowledge of world ethnography than any anthropologist I have ever known."[5]

In between sojourns at the farm and in the library, however, military events took him to Mexico in 1916 as a national guardsman under Pershing to quell Pancho Villa's uprising, to World War I campaigns as an artillery lieutenant, and to the West Pacific in World War II as a lieutenant colonel doing anthropology and police work. Throughout his life, his armchair regimen was interrupted by more regimental activities. (He also volunteered his services to J. Edgar Hoover, informing on the leftish tendencies of his anthropological friends and acquaintances during the 1950s, a fact, only just released from FBI files, that has recently taken on the flavor of an exposé in anthropological circles.[6])

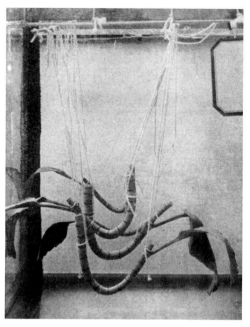

Jacques Loeb's demonstration of geotropism at work in *Bryophyllum calycinum*: "These stems were originally straight and suspended in horizontal positions. In about ten days they bent, becoming concave on the upper side." (From Jacques Loeb, *Forced Movements, Tropisms, and Animal Conduct,* Philadelphia: J. B. Lippincott Co., 1918)

Rheotropism, as demonstrated by Loeb: "Influence of motion of the hand of an observer on the direction of the motion of a swarm of sticklebacks in an aquarium. The arrows indicate the direction in which the hand was moved. The swarm of fish moves always in the opposite direction in which the hand is moved." (From Loeb, *Forced Movements*)

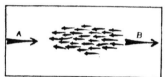

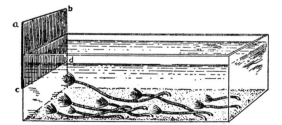

Heliotropism, the most common tropism. Tube worms in the aquarium are all bending toward the light. (From Loeb, *Forced Movements*)

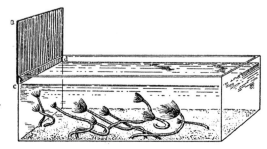

The same animals after the orientation of the aquarium toward the window was reversed. (From Loeb, *Forced Movements*)

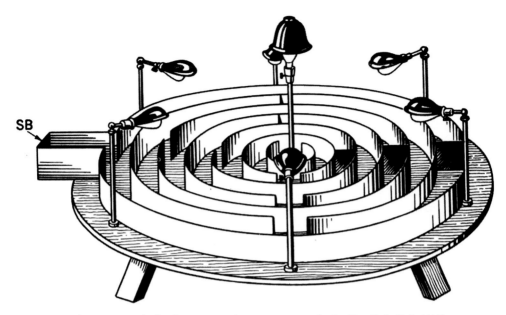

SB

John B. Watson's circular maze. (After J. B. Watson, *Behavior*, New York: Holt, 1914)

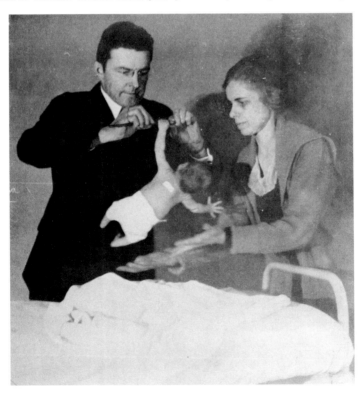

Watson tested baby reflexes in experiments with human infants
from 1916 through 1920. Here he tests the tonic grasp reflex.
(Courtesy Johns Hopkins University Archives)

Examples of the fast-growing and increasingly ingenious designs for rat-in-maze laboratory research during the 1920s and 1930s: here, a maze for testing visual cues; a "double alternation tridimensional spatial maze"; and a "narrow path elevated maze with T-shaped units." (Images courtesy of American Psychological Association and Heldref Publications)

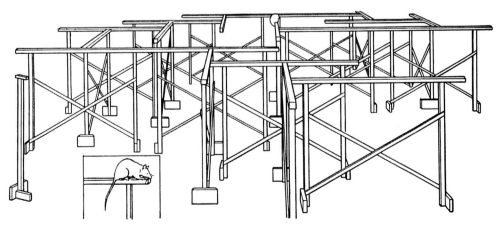

The Institute of Human Relations at Yale, built in 1929 with the help of $7.5 million from the Rockefeller Foundation. (Courtesy Yale University Manuscripts and Archives)

Clark Leonard Hull galvanized a young group of experimentalists at Yale with his goal of creating a unified theory of human behavior. Eventually he developed eighteen theorems in his 1943 *Principles of Human Behavior*, and left behind seventy-three "Idea Books" full of designs for thinking automata, or "psychic machines." (Courtesy Yale University Manuscripts and Archives)

John Dollard, a psychologically inclined sociologist and the author of *Caste and Class in a Southern Town*, an American classic. He played a pivotal role at Yale in reconciling behaviorist science with Freudian psychoanalysis. (Courtesy Yale University Manuscripts and Archives)

Neal Miller, shown (*left*) during his early Yale years and (*right*) with laboratory machines, was an experimental psychologist who partnered in research with John Dollard and helped advance the influential frustration-aggression hypothesis. (Courtesy Yale University Manuscripts and Archives)

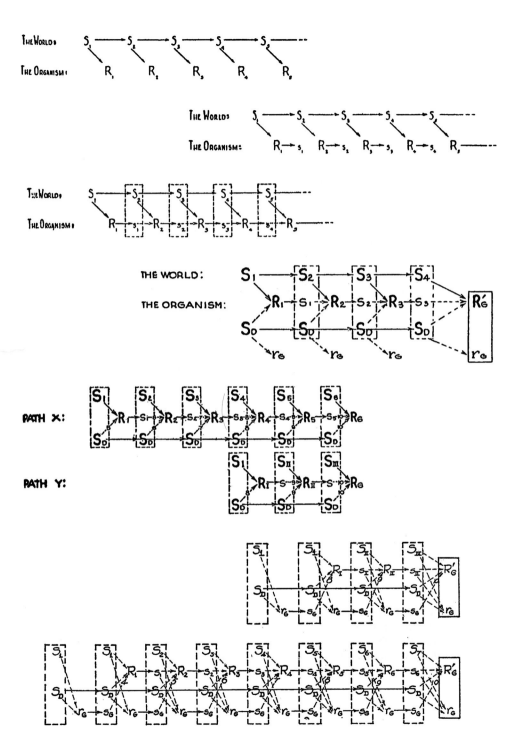

Stimulus-response mechanisms depicted in diagrams from Clark Hull's articles in scientific journals. From 1929 to 1936 the mechanisms became increasingly complex, like fine-spun webs. These diagrams ultimately are intended to represent such things as "knowledge" and "purpose" in graphic form. (Courtesy American Psychological Association)

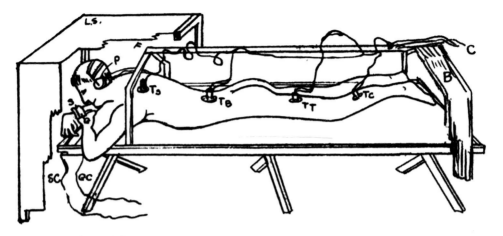

Conditioned fear reaction in human subjects. Illustration of an experiment run by Hull at Yale on university students in 1937: "The subject lay completely nude on an ordinary army cot . . ."
(Courtesy American Psychological Association)

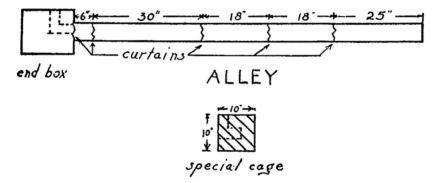

Design of a 1934 experiment at the Institute of Human Relations by Neal Miller, in which highly "motivated" rats (i.e., rats kept very hungry and thirsty) were trained to run down an alley, through some curtains, and into an "end box" where, in order to get food and water, they were forced to turn left or right in a "special reward device."
(Courtesy American Psychological Association)

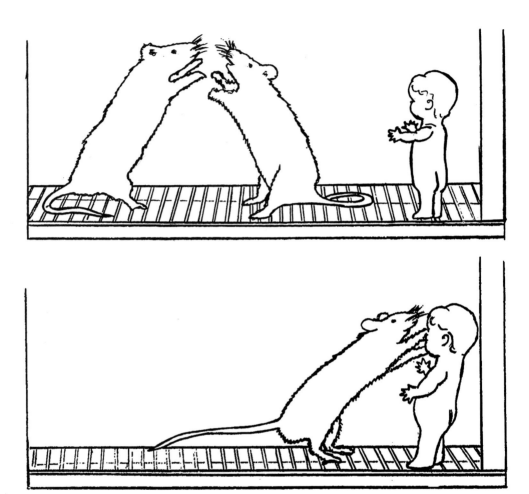

Rats at the Institute of Human Relations were electrically shocked on a charged wire grid until they displayed "fighting" behavior. They had to "attack" each other for the current to stop. Soon they attacked each other automatically as soon as they were placed on the grid, even with the current off. Next, one of the rats was removed and a celluloid doll put in the cage. Now the rat "attacked" the doll instead. This was seen as an experimentally created version of scapegoating and other human responses in which frustration generates aggression and that response is transferred—"generalized"—to another object.

(Courtesy American Psychological Association)

In the spirit of Murdock's career-long conviction of the key importance of kinship charts, it is tempting to focus on the father-son dyad. (In his only published autobiographical account he did not mention his mother except to note her pristine but penniless old New England heritage.) A gentleman farmer whose New England stock was apparently so rooted in the soil that he never saw fit to go to college, Murdock's father did not let his lack of higher education prevent him from attaining thorough and unrelenting convictions on politics (pro-democratic) and religion (anti-all). As his son recalled, he "felt very strongly that it [the church] was superfluous for people of character and unacceptable on principle for people of intelligence and education."[7] His son violated this principle when, wishing to take greater advantage of their clergyman's excellent tennis game, he formally joined the church for a short period in his youth—a move that greatly disappointed his father but seems at least to have paid off tennis-wise, since Murdock made it to the Forest Hills tournament during college. Like many male members of his family aside from his father, Murdock attended Yale as an undergraduate, which was "easy and pleasant" for him, as the rest of his life in its broad outlines promised to be for one in his position. He chose Harvard for law school.

These easy and pleasant expectations changed abruptly when his father died in 1920. Murdock dropped out of school and, with the help of a $50,000 inheritance, took off to travel extensively in the Orient, visiting Japan, Korea, China, Malaya, Indonesia, and India. This crisis was the immediate spur to his becoming an anthropologist. After Murdock sought and failed to receive admission to graduate school at Columbia, he tried Yale. There he succeeded, studied with the old-line evolutionary sociologist Alfred Keller, and was shortly thereafter appointed to the department of sociology and anthropology.

The first recorded mention of an ambitious filing cabinet of facts was young Professor Murdock's effort, beginning in 1928, to compile a bibliography of all known cultures. At this point he envisioned the files as a pet project and solo venture. In 1931 he described the files as "a comprehensive study of the cultural traits in 2000 primitive tribes. The problem is to test the various theories of social evolution by statistical techniques."[8] At this early stage, then, the files were very much in the nineteenth-century spirit of looking for clues to the evolution of types of human societies, making use of twentieth-century techniques of collation and statistics. Within a couple of years, however, Murdock and his files were successfully drafted into a cutting-edge program of social and human engineering, after which they seemed to take on a life of their own.

IN ORDER TO ACCOMPLISH the admittedly enormous task of filing "all aspects of human existence" among each of four hundred ethnic groups (figured as approximately ten percent of all known cultures), the files had to be refigured as a team project. Five graduate students and young professors in anthropology and sociology convened under Murdock to discuss its carrying out, including Clellan Ford, Alfred E. Hudson, Raymond Kennedy, Leo Simmons, and John Whiting, with a team of other graduate students beneath them to "process" the information, and an auxiliary team of typists, office workers, and graduate students' wives to do the clerical work. It resembled a factory assembly line, except that the work was to take apart books and distill them down to standardized facts, rather than mass-produce objects out of parts.

The head team of six agreed that the best way to start would be to split cultures into their component bits and likewise divide the world into its component parts. A classifying scheme for the totality of culture, eventually published as the *Outline of Cultural Materials*, along with a classifying scheme for the totality of societies, the *Outline of World Cultures*, served as indexes to the files. The latter (the world) was relatively easy to carve up and designate: each geographical area was assigned a set of drawers, beginning with Asia in the upper left-hand corner, so that Iroquois became "NM9 Iroquois," and the Admiralty Islands "OM6 Manus." Dividing up the former (culture itself) was more of a challenge. Although the participants were aware of their predecessors' attempts at such schemes, none of these was felt to be either standardized or comprehensive enough. Instead, the team would start from scratch, assigning to each member several "large blocks" of culture, such as Kinship, Magic, Politics, the Reproductive Cycle, or Material Culture and Technology, each of which would be divided into a taxonomy of the most logical possible sort.

After experimenting with different divisions, each scholar reported back with his conclusions on the most seemly breakdown. Two-digit numbers from 10 to 88 marked each major heading and a third or fourth digit marked each subdivision thereof, thus communicating with that numerical coding a confidence in the impartiality and neutrality of the divisions provided. The two-digit categories would typically be broken down into further distinctions marked by three-digit and even four-digit codes. (Other codes were adopted as well—bracketing, asterisking, zeroing, and superscripting marks, which when used freely give an arcane and hieroglyphic feel to some of the entries.)

Still, a "textual" issue beset Murdock's team: to what extent should the original texts be allowed to survive the top-down filing process intact? Many of the sources to be filed were written not by scientifically trained observers but by travelers, missionaries, journalists, "indigenes," historians, and "uncategorizable or unknown" others. Yet despite their sometimes ragtag origins, the sources held information that would have to fit the neutral *OCM* categories. Some argued that only the data itself, perhaps reduced to binary code form, should be culled from the ethnographic sources, while others argued that certain relevant excerpts of the texts should survive entire. This was Murdock's position, and it won out: "So it was decided that whatever was done— even though the information might be characterized and classified into pigeonholes, as it were—would be done in the original text. . . . In other words, this would be an organization and ordering of original information and not simply a[n] . . . abstracting or coding system," recalled one participant.[9] One reason for this decision may have been that the texts were often the work of highly peculiar individuals and it would be better to preserve their eccentricities. Another may have been Murdock's feeling that by preserving the "original information," the files would be more authentic. At any rate, within each file, they decided, "the standard unit of analysis is the paragraph."[10] From countless ethnographic, visionary, and morally suasive texts—duly processed—would come the humble yet somehow perduring standard units that would be used to build the files.

The group devised a method to "process" the texts. Any relevant sentences or fragments of sentences were marked by a processor (usually a male graduate student), then transcribed from the source, typed on a five-by-eight-inch page, and filed according to the appropriate geographical area and topical two-, three-, or four-digit heading (usually by a female secretary, female student, or wife of a graduate student). In order that the bowdlerized excerpts did not float free of their original sense, a copy of the paragraphs of origin could be consulted in their entirety, save for certain parts deemed irrelevant, under category 116, "Texts."

Under AU1 Afghanistan, *OCM* category 281, for example, is an excerpt from a contemporary text titled "Travels in Afghanistan." The top line of the index card reads like a code, giving its author identification as N (for Natural or Physical Scientist), its quality as 3 (good), and its cross-references to categories 291, 281, 443, 609, 275, 326. Below this line, the excerpt reads, "In the Kabul-Ghazni area sheepskin poshtins are the greatcoats worn in winter. They are made of soft-dry-tanned pelts, like chamois." Turning to category 116, "Texts," one finds a fuller excerpt, where, following several cataloged paragraphs describing in detail the workings of the Kabul-Ghazni sheep market, there is a passage marked 000:

> And then we came again to the entrance where, dazzled by the sun, we groped our way toward our car.
>
> One comes away from a first visit to this ancient market place as from a world apart; it is a spot as yet untouched by Western influence. And may nothing ever change it! There is more real contentment on the faces that pass these busy crossroads of Afghanistan than in all the hurried crowds that pass by Piccadilly Circus or Times Square.

The mark 000 tells the investigator that the information—the sun dazzling, the market busy yet somehow unsullied by hurrying New

York– or London-style crowds, the assertion of a greater sum of "real contentment" found there—is not relevant. The next paragraph resumes providing file-worthy information: 326, "I sent for a local black smith to build a baggage rack on top of my car." In other cases, the files' processors would excise such commentary. Critics of the files, from their earliest incarnation on, would insist that despite the triumph of "original information" over pure data, the processing method was a weak point and did violence to the texts used. For these critics, the welter of codes, categories, and marks revealed a lack of logic presented as supremely logical, and "the idea of breaking documents up into pieces and shuffling them arbitrarily, as it were, and out of context" was anathema.[11]

During this time the filers and their project came into contact with a program at the nearby Yale Institute of Human Relations. There experimental social scientists were turning to the field of anthropology to help build a unified theory of human behavior. With the greatest access to the world as a whole, anthropologists were equipped to bring in its data.

Not just any anthropology would do, however. Many of the ablest anthropologists had portrayed some cultures' moody variability and just plain strangeness—Marcel Mauss in France liked to speak rather poetically of these "dead moons" in the "firmament of reason." For some, the more they scrutinized a strange culture, the stranger it became. There was something surreal and jarring about the ethnographer's contact with different cultures, a clash of perspectives that could not easily be smoothed over, and for some during the 1930s, as James Clifford has remarked, "the exotic was a primary court of appeal against the rational, the beautiful, the normal of the West."[12]

Murdock and the institute experimentalists did not care to make such an appeal, and for them the manifest strangeness of culture suggested on the one hand a threat of chaos and unknowability, but on the other—in the right hands—an ideal testing ground for the universal, a perfect range of variability, a data bank of facts that might make, by their sheer mass and statistical arrangement, a case for the normal.

They needed a type of anthropology that could present culture as a "perfect pattern from one generation to the next."[13]

The marriage of the files to the institute lasted only about five years, but it helped the files flourish and find their footing—so that ultimately, as the result of emergencies of war and accidents of fate, the files outlasted and outflourished the institute itself. By 1940, with the filing system in place, Murdock's team had "processed" almost one hundred cultures and duly filed the resulting cache of index cards. Enthusiasm was high, and a sense of mission—in which anthropology contributed to the greater good of a grand theory of human behavior—pervaded the project. The files took the offerings of anthropological field workers—too often random in presentation and inscrutable in argument—and rendered them as parts of an orderly system. The timing of the initial filing was perfect for the outbreak of war, and when Murdock and his team were deployed after Pearl Harbor to little-known areas of the world, "the files . . . acquired a practical value which had not been anticipated," as they put it with some understatement.[14]

WITH THE BOMBING OF PEARL HARBOR, Yale's campus, along with the rest of the nation, went into upheaval. Students rioted spontaneously, trashing the lobby of the Taft, the grandest hotel in town, and took up arms against the New Haven police. In short order the campus turned itself over to the military; as *Newsweek* announced, "Yale Blue has gone to war." A Committee on Student Preparation for War Service evaluated each undergraduate course in the curriculum for its relevance to the war effort. The army air force took over half the living quarters and a third of the rest of Yale's plant. Navy, marine, and air force reserves were also in place. More broadly, Yale announced that it would "seek to devise means of closing the gap which exists between college and the armed forces" so as to make its students more ready soldiers.[15] As the historian Robin Winks described the atmosphere:

> By the summer of 1943 Yale had become a military camp. . . . Seated and served meals in the dining halls, with silver and crisp napery, gave way to

standardized trays and cafeteria service. . . . Popular lecturers, accustomed to having students hang on their every word as they built to resounding conclusions, normally to halt dramatically at precisely the fiftieth minute, found their perorations rudely interrupted by military marching bands.[16]

Surprisingly enough, in the heat of war preparations the most useful of Yale's going concerns was not psychological or biological but anthropological, in the form of Murdock's files. Even those who were skeptical about the work of the institute (not to mention its members' unflagging efforts to publicize that work) eventually acknowledged the key role the files came to play in Latin America and the Pacific campaign.

With the war, quick methods and full coverage were indispensable, and whereas other scholars had to gear themselves and their skills to these imperatives, the files had already been built according to those imperatives. As Barry Katz points out, "One of the early achievements of th[e] first generation of American intelligence analysts was to demonstrate that it was possible to secure the greater part of this vital intelligence not by dropping behind enemy lines but by walking over to the Library of Congress, where they did what scholars do best, namely, plodding through journals, monographs, foreign newspapers, and other published sources."[17] This was much easier for Murdock and his team who, unlike the esteemed Frankfurt School historians to whom Katz refers, had already built a bureaucratically structured machine for translating, culling, expurgating, distilling, filing, and cross-referencing such information. With 145 so-called "backward cultures" processed, many trips to the library had already been spared. This fact-processing may have taken place assembly-line style along anti-intellectual lines, reduced the role of the scholar to processor, and resulted in a low-grade product at times, but the system would be all the more easily assimilated by military and government concerns.

This system, along with Murdock's personal qualities, impressed army and navy intelligence, which was often skeptical of professorial types such as émigré scholars Herbert Marcuse, Franz Neumann, and Otto Kirchheimer, who joined up and were dubbed "the Bad Eyes

Brigade." Unencumbered by foreign accents, ethnic origins, aversion to military training, or Hegelian-style prose, Murdock and his team found favor among military intelligence and even military operations groups everywhere from the State Department to the army and navy. Whereas most professors who "joined up" remained stateside or, at the most, were dispatched to a foreign branch of a research unit, Murdock and two of his disciples joined Admiral Chester W. Nimitz's invading fleet and saw active duty in the Pacific. And unlike any other professor known to my researches at least, Murdock—soon to become Lieutenant Commander Murdock—actually served as a police officer in military affairs, not just civil affairs, at Okinawa in the final months of the war.

THE OFFICE OF THE COORDINATOR of Inter-American Affairs under Nelson Rockefeller commissioned Murdock's group to build the Strategic Index of the Other Americas (later renamed the Strategic Index of Latin America, or SILA), "to bring together in one great reference center all information on all the other American Republics as would be of utility to the Government, most especially the armed forces, in connection with the war effort." The government and military found they were "profoundly handicapped" by the lack of such an effective information system and were delighted to discover the Yale files.[18]

One person who felt the Yale files might redress this handicap was the chief of the OSS's Latin American section, Preston E. James, who championed Murdock. He probably had Murdock's files in mind when he criticized other academics working in Washington who, he said, "are disdainful of the encyclopedic approach and who insist that the mere gathering of facts is a useless occupation."[19] By eloquent contrast, at the files "mere" facts were the order of the day and the encyclopedic approach unchallenged. The Yale institute's director turned over fourteen offices on its top floor (formerly the home of some of Robert Yerkes's primates) to Murdock's expanded filing team so that, with the aid of abstractors, file clerks, secretaries, typists, a photographer, a draftsman, a bibliographer, two full-time bookkeepers, and a newly hired corps of translators, they could press on with processing and filing at peak levels. By November 1942 Murdock announced with some satisfaction that 320 English sources and 229 foreign sources, for a total of 30,467 pages on Latin America, had been filed, with more coming down the pike. He predicted that his Strategic Index would go far to "eliminate repetitive and time-consuming library research" on Latin America, a prospect government specialists found enticing: "By this method anyone can find in a few moments *all* existing material on every vital subject."[20] (Subjects vital to the State Department included the ways and mores of the Amazonian peoples, who unwillingly worked the rubber plantations of the U.S. Rubber Reserves and Fordlandia. Rubber, as a cash crop, had widely replaced subsistence crops in the Amazon Valley, and by the summer of 1942 a quarter-million native workers and their families suffered from malnourishment at emergency levels. Murdock and his team used information from the files for their reports, among them "Potential Indian Labor Supply in the Amazon Basin," "Food and Food Habits of the Natives in the Amazon Basin," "The Way of Life of the Rubber Gatherer with Special Reference to His Food Supply," and "The Preserved Foods of the Aboriginal People of South America.")

As the Pacific campaign intensified, the navy declared Murdock's work on that area "of immediate military value," so that in March 1943 Murdock accepted a commission as a naval lieutenant commander. In July his files were drafted wholesale as a navy research unit on Oceania. Murdock brought along two of his younger anthropologist colleagues, Clellan Ford and John Whiting, and both men accepted navy commissions as lieutenants junior grade.

As the American fleet island-hopped its way across the Pacific, occupying each tiny island in a sea of tiny islands, the military government, or "Milgovt," people looked to Murdock's team for vital information. The files helped answer questions such as *Where are the best beaches and water sources? How do you eat a coconut? Which of the locals may be friendly to the Japanese?* and *When should you pat a native on the head?* (Never.) Cultural and psychological data culled from the files helped in "concentrating" natives in camps—work carried out by anthropologists and other administrative officers.

By June 1944 Murdock and his two colleagues had left the Marshalls and were on their way to active duty at Okinawa. His letters from the front were unusually ebullient. On July 29, 1945, he wrote to the institute's director that he had been doing police work for the past two months in an unsecured area, "where we were repeatedly subject to Jap attack and ambush. . . . [T]he situation was exciting, my work was productive, and I really had the time of my life." He also mused on how his earlier work at the institute was panning out: "My anthropological and Institute experience have [sic] proved most useful in practical matters." Out of their experiences in Okinawa came, for Murdock and his team, a new commitment to the files: "We all want to . . . emphasiz[e] [the files'] use as a testing ground for scientific theory rather than the mere further accumulation of materials," he wrote from the front. They were confident that the files' usefulness in war promised future usefulness in science. As a result of military support for expanded operations, the bank of data was full and "I want to see some checks drawn on it before putting anything else in," Murdock wrote.[21] He recommitted himself to making a true science out of social science, on a scale that only the government or navy could fund.

The files, conceived at one time as an old-fashioned encyclopedic device and sometime later as a mere tool to aid the institute's grand design, now flourished. Scientists presented the expurgated bits of texts they collected, shorn of their original sources, as solid and unimpeachable edifices of fact. Paradoxically, the more they claimed the facts were unchanging, the more they displayed their susceptibility to alteration. So it was that the files were "markedly influenced" by the war, elevated to the status of science at the same time they were celebrated for their "potentially great practical value" for further intelligence, military, and governmental purposes, a value now hard for their creators to ignore.[22] The navy, army, air force, and Central Intelligence Agency gave $50,000 each for research concentrating on Southeast Asia, Europe, Northeast Asia, and the Near and Middle East. Their purpose was to use the files to help administer geographical hot spots.

Crowing over the money pouring in from the government, the files' authors struck a new note of entitlement. According to Murdock's protégé Ford, the government "could scarcely afford not to support an organization that could supply it with accurate, critically evaluated, usefully organized, basic information on peoples of the world."[23] Meanwhile rival anthropologist Clyde Kluckhohn worried that Murdock and his ilk were "possibly a trifle intoxicated by the fact that for the first time men of affairs are seeking [their] advice on a fairly extended scale."[24] No longer simply the humble bearers of academic theories about universal laws, the files were newly defined by their administrators' pride of service. To the extent they were practical, they were scientific, and by their mere existence they were framed as honorable.

Their usefulness also marked the arrival to full acceptability of the behavioral sciences, which soon became humming nodes of activity at home and in cold war forays into faraway places. Such scientists' special knowledge of the world and its formerly inscrutable corners, and of human behavior and its multiform intricacies, led to new vistas. Government officials ascribed to behavioral scientists an unnameable but certain potency, which coincided with the general tendency in the postwar years to think a great deal of experts and their expertise. Few rose higher or were better listened to than the file builders.

THE FILES WERE A GREAT AID in evaluating the world during wartime, and they also helped fit the American way of life into a necessary new perspective. Three of the most significant books to come directly out of the file project—Murdock's 1949 *Social Structure*, Ford and Beach's 1951 *Patterns of Sexual Behavior*, and Whiting and Child's 1953 *Child Training and Personality*—addressed the question of how America and typical American practices compared with those of the rest of the world. The aim was normalization: to define, support, calculate, recalibrate, and reinforce the norms by which people should live. The books, which respectively addressed the domains of sex, family, and childrearing, collectively mapped the outlines of the domestic front and used the files to marshal the information needed to manage lives and lifestyles.

Ford and Beach's volume came out three years after the publication of the Kinsey report, *Sexual Behavior in the Human Male*, and was introduced in relation to that project. Displaying the more-universalist-than-thou aspirations typical of the Murdockian file-gatherers, Ford and Beach claimed to have outdone Kinsey's team by providing thoroughgoing data on sexual behavior in not one but 190 human societies, as well as a range of warm-blooded animal species. They found the Kinsey report commendable but "not in itself an adequate basis for a comprehensive understanding of human sexuality," and they believed that only when sexual behavior was examined within the full scale of its variation would true scientific understanding result. In the case of the human species, the authors pointed out, such a scale was best compiled through the use of the Yale files, "instead of years of research."[25]

This cross-cultural and cross-species perspective, they felt, amounted to already-performed "natural experiments." A sense of possibility pervaded: "If one wishes to learn the marital effects of socially approved premarital sexual freedom upon marital fidelity, he cannot set up an experiment in our own society to find the answer, but he can compare the behavior of people in cultures that permit some sexual liberty before marriage and in those that do not." Gathering an immense range of facts and fathering a massive information system was a way of conducting implicit comparative experiments. It also had the advantage of being both scientifically neutral and morally commendable, thus explaining how the authors could claim at once a "self-imposed avoidance of value judgments" in their work and a "moralistic" purpose in their work of facilitating judgment.

Through such experiments, the social consequences of any hypothetical permutation of sexual behavior could be gauged—for surely it, whatever "it" was, no matter how strange, existed somewhere on this vast scale of behavior. And if it could be located, it could be measured and judged on the basis of its measure.

This suggestion of the malleability of behavior on a large social scale dominates Ford and Beach's book; yet at some points the book seems also to function as a behavioral guide for small-scale individual behavior, reassuring the reader, for example, that although face-to-face copulation with the male above the female is undoubtedly the most common position among humans of varying social groups, a fair number of other positions should be considered normal in their contexts and may, in point of fact, give more pleasure to the woman. An interesting tension arises: on the one hand, covering a comprehensive range of species and societies raises the possibility of making any deviation normal; on the other, the comprehensive range ends up circling around the concerns of the middle-class, educated, WASP American male. The book includes a guide to the exact layout of the female anatomy, with relative positioning helpfully given "if the woman is lying on her back." A chart of the near-infinite sexual variety of humankind also turns out, on closer scrutiny, to be a road map of the sexual specifics of the mid-twentieth-century man.

Likewise in Murdock's 1949 *Social Structure*, his major book, the domains of family and kinship, after receiving thorough and wide-ranging inspection, point to concerns specific to the Connecticut Yankee. Murdock gathered data from 250 cultural groups, using his files for 85 of them and library research for 165 more. He then applied the statistical method and Yule's coefficient to these data. Although the "Connecticut Yankee" happened to be only one group of many, counting for a single unit of analysis on a par with, for instance, the Yoruba, this statistical parity does not hold.

Much of the book is taken up with definitions of kinship forms; this was Murdock's forte, and he was widely respected on both sides of the Atlantic for it. (Even in Oxbridge anthropological circles, where kinship studies were pursued with great vigor and exactitude to the exclusion of almost all other concerns, he was held in high regard.) By the liberal use of what one reviewer called "Professor Murdock's terminological hyper-precision" in doling out helpful corrections to one scholar's description of a kinship system, necessary additions to another's, Murdock gives the impression of being utterly comprehensive and evenhandedly expert.[26] Yet one of his central claims—that the nuclear family is "a universal social grouping . . . [and] exists as a functional group in every known society"—was not true. At that time, ethnographic research was demonstrating ever more conclusively that the nuclear family was not the basic unit in many Asian and Melanesian societies; the only way to conclude that it was universal was to imply that the mere biological existence of a mother, father, and siblings constituted a social form. Murdock's assertions, for example, that "in 140 out of the 187 in our sample for which data are available—nuclear families are aggregated, as it were, into molecules" and "clusters of two, three, or more are united into larger familial groups," send up a smokescreen of confident precision even as they undermine the argument itself.[27]

Murdock defines the community as "the maximal group of persons who normally reside together in face-to-face association." Along with the nuclear family, he considers this group "genuinely universal." This is a roundabout way of saying that humans live in groups just about everywhere, but Murdock goes on to make great claims for this truism and argues that these groups not only are universal but function as "the primary seat of social control." They socialize individuals and channelize behavior. He seems to suggest that a mechanism for social control is built in to all human groups (in the form of the family and the community), and that if one could access this seat of control, it could be used to regulate behavior.

Like Ford and Beach, Murdock concludes that all forms of behavior, even deviant ones, can potentially be made normal. For example, despite the Mormon failure with polygyny, he suggests, there is "perfectly clear" ethnographic evidence "that it can be made to work smoothly." Underlying such statements is a concern with the inevitability of change, a reckoning with the undeniable fact that "fundamental change does occur." In an unstable world, anything can be modified, and Murdock's implicit interest was in smooth adjustment and conservative control. His last chapter offers a "universal law of sexual choice," which suggests (after the elucidation of seven "gradients" affecting this choice) that "our own particular social structure predisposes the unmarried American male to prefer, both in marriage and in informal sex liaisons, a woman of his own age or slightly younger, with typically feminine characteristics, who is unmarried, resides in his own neighborhood or at least in his own town, belongs to his own caste and social class, and exhibits no alien cultural traits." The universal law, it seems, ensures that American men prefer the girl next door. A towering normality emerges from this disquisition on the universal. (As if to offset the overstimulation his law might arouse, Murdock observes that the preferable "sexual object"—having been selected by the identical standards—is a man's wife.)

The argument seems esoteric today. To paraphrase it: the nuclear family is very important and is at the root of most other social arrangements. In light of Murdock's lengthy elaborations of the patently obvious, it is strange that the book was so well received at the time. Even the British anthropologist Sir Edmund Leach, after pointing out the banality of Murdock's conclusions, which "makes one wonder whether all this pother about kinship really has any significance at all," commended the book as "brilliant, penetrating, and provocative." Similarly, another reviewer noted that "this dressing up in pompous terms of the wisdom of our grandmothers . . . sometimes makes one wonder if there is not a positive gradient of banality in the work of social scientists" and yet also maintained it was an "important volume."[28] The mix of self-evident conclusions, myriad statistically correlated charts, and triumphant claims—such as Murdock's proof showing that anthropological data on kinship terminology, as expressed in 26 theorems and 155 computations, is "as susceptible to exact scientific treatment as are the facts of the physical and biological sciences," and thus constitutes a result "perhaps unprecedented in social science"—makes for an argument remarkable on many levels, including the seriousness with which it was treated.

Whiting and Child's extremely influential volume *Child Training and Personality* set the standard for postwar culture and personality studies. Like the two volumes already discussed, this study came directly out of work at the institute with the files and was guided by a common theoretical design that the authors characterize as a mix of modern behavior theory, cultural anthropology, and psychoanalysis. The main influences cited were Murdock for his cross-cultural method of taking the single culture as the unit of analysis; Ford for his influence on the derivation of basic concepts; and Dollard, Hull, Miller, Zinn, Mowrer, Doob, and Sears, all core members of the Yale institute, for their reconciliation of psychoanalysis with general behavior theory.

Of the seventy-five primitive cultures in the sample, sixty-five came from the files. The authors chose five dimensions universal to human behavior—oral, anal, sexual, dependent, and aggressive—and measured child-training practices across a range of primitive societies. These five dimensions were assumed to be universal due to infants' universal experience of helplessness: "In all societies the helpless infant, getting his food by nursing at his mother's breast and, having digested it, freely evacuating the waste products, exploring his genitals, biting and kicking at will, must be changed into a responsible adult obeying the rules of his society."[29] This was a derivation of Freud's argument in *Civilization and Its Discontents*, but instead of relying on evidence plumbed from the depths of the unconscious, Whiting and Child relied on data pulled from ethnographic extracts filed in their cabinets. Under the heading of "Infancy and Child Care," for example, one finds six categories, including 485, "Infant Care," and 486, "Infant Development," along with many subcategories and cross-references. Subcategory 4841 lists "ideas about breasts" and is cross-referenced to 4519, "Human Organism; Torso."

Although the authors insist that they restricted their focus exclusively to primitives to avoid getting mired in the undue complexities of modern European-American societies, a key chapter of the book locates Euro-American middle-class culture in relation to the array of primitive cultures. For each of five domains of child training, the most and least extreme variants were given and compared with data from "50 middle-class families living in Chicago in the early forties." They did this in order to guide the reader with "comparative information," should he "wish to consider the probable wisdom of any specific change that is suggested" for his own society.

In the first area, "nursing and weaning," fifty-odd primitive cultures and a single middle-class culture were evaluated by judges and rated on a scale of 1 through 18 for their weaning customs (1 represented an indulgent type of weaning that let the child gradually stop on his own initiative and 18 was considered an abrupt and severe cessation). The American middle class was "found to be less indulgent than most primitive societies" and ranked second only to the extremely strict Marquesans in severity. In the second area, "anal training"—that is, toilet training—the Americans ranked as less extreme than only the Tanala of Madagascar. Likewise in the third area, "sex training," which includes separate ratings for masturbation, heterosexual play, immodesty, and overall sexual behavior including homosexuality, the Americans received ratings ranging from third-most-severe (after the Tanala and the Ontong-Javanese) on homosexuality, to second-most-severe (after the Manus of New Guinea) on masturbation and its consequent social anxiety. In the fourth area, "dependence training," the Americans registered at two points below the median: an early age for independence in comparison to most primitives, "but not aberrantly early." In conclusion: the American middle class, although on the extreme end of a range of severity, "is not outside the range of primitive societies," tied overall as it is with the two strictest primitive groups. Notable, again, is the exquisitely explored range of variation, the marking of deviant childrearing habits, and the redefinition of them as merely "extreme"—an inherently relative term. One can be fully normal only in relation to something or someone that is not.

By introducing human judges into the process, the authors worried that subjectivity would intrude: "Could it be that the judges agreed with each other because they all knew what we wanted or because they worked together?" Elaborate further precautions were taken to avoid such dangers. Two out of three judges were chosen in part because they had never worked with Whiting and Child before, and all three were sequestered from prejudicial theoretical information while working as judges. As a further guarantor of nonsubjective judgments, the judgments of any two judges were periodically statistically correlated. In this way a strikingly thorough series of steps were taken to ensure impartiality, and the authors noted that even if "some readers may feel that the measures we have taken to insure against contamination of the judgments are rather extreme," they were successful in their aim: "It was thus virtually impossible that the judgments could have been contaminated by the judges' wishes or preconceptions about the outcome of this research." More urgent epistemological issues concerning the reliability of the files' textual extracts, not to mention the "contamination" of living in mid-twentieth-century American society, were neatly sidestepped. Like their teacher, Murdock's students resorted to elaborate, quasi-scientific devices that they hailed with great fanfare, in order to forestall the obvious and simple problems inherent in this kind of research: How to quantify the unquantifiable? How to bring the vagaries of human existence under scientific scrutiny and control?

BEFORE THE WAR, Murdock and his team had not always been successful in turning texts into neat, hygienic piles of facts, but during the war, the files added up to a consensus, a working thesis, a bottom line made of facts themselves, on which the navy's invading fleet or the State Department's official functionaries could gamble. Even the mobilization of pencil-pushers and fact-finders, the social scientists and historians who made up the "chair-borne divisions," were a part of this expansion of scale. It was not the mere temptation of power but a vein of the heroic that wooed them. In the process, as an unanticipated consequence, the fact-finders began to transform the relation of the rest of the world to the United States.

The Yale files were an ambitious project, and even the graduate student of Murdock's who eventually took them over recalled that its creators worried at the outset that the venture would be "too much to cope with."[30] How they coped, specifically how these scholar-scientist-administrators undertook to organize and build a proto–World Wide Web, a box of the world, a limitless incorporated file, is one side of the story. For what purpose they coped is the other. How did the files help shape the terms of a debate in society at large about the role of science in relation to the hitherto private psychological and personal dimensions of social life? How did the files become the instruments of (to use the sociologist Erving Goffman's phrase) a most deliberate "bureaucratization of the spirit"?

CHAPTER 9

Anthropology's Laboratory

AT THE MOMENT of the United States' accession to number-one-world-power status, just as the Pacific campaign of World War II was being fought and won, pivotal events took place on islands in the middle of nowhere. The Marshalls, the Carolines, and the Marianas were generally described as "far-flung," as if they were suffering the effects of some careless invisible hand. Eisenhower's 1947 estimation of them—"Nothing but sandpits"—suggested utter irrelevance. Yet it was here, in these newly occupied areas, that a baton passed from one set of rulers to another: Americans took over what had been, since World War I's Versailles Treaty, part of Japan's Greater East Asia Co-Prosperity Sphere, and to devise policy for civil affairs in the Japanese Mandated Territories, they largely looked to the social sciences. Human engineering principles and techniques paved the way for anthropologists and sociologists to help the military government remake life on these islands.

Americans forged a new style of occupation. Murdockian anthropologists and other social scientists played roles, large and small, in passing from a variously British, Japanese, or German type of colonialism to a new sort of governance—the navy's Program for Occupied Areas. The occupation of these Micronesian atolls took on a distinctive "social science" character and was like an ongoing set of experiments—all of which coincided, as it happened, with the largest research initiative in the history of American anthropology, and the most massive tonnage of nuclear test bombs ever dropped on a single area. (Between 1946 and 1958, sixty-eight nuclear bomb and missile tests were performed in the Micronesian area of the western Pacific, earning it a central place in what policy documents would come to call the "Nuclear Pacific.") The islands of Micronesia became a living laboratory.

The events that took place on these small but strategically important spots are mainly forgotten, if they were ever known. They have interested few observers save those who live there (or in the case of Bikini, who used to live there). Kapingamarangai, Jaluit, Kusaie, Likiep, Truk, and Yap are a few of the ninety-eight islands or island clusters spread out over five million miles of ocean.[1] Most discussions of American occupations during the Second World War focus on Japan and Europe, for obvious reasons of scale and geopolitics. And although U.S. rule in Japan and Germany was also characterized by an experimental social science approach, it had considerably freer rein in Micronesia.

In general, the violence during the war was intense, and so too were the various waves of war relief, matériel, internment, administration, management, and mutual adjustment that followed. Yale-trained and Rockefeller-groomed scientists approached these movements, before and after, with a style of anthropology that was geared to serve the unified social sciences and the larger project of human engineering.

CUT TO THE SECOND WORLD WAR, Pacific Theater: as the war against Japan intensified and invasion after invasion of new territory was planned, it became resoundingly clear that the United States lacked a single school or training program for running an occupied area. (The only U.S. possession at the time was the Panama Canal Zone.) Committees of internationalists and experts convened to discuss training officers who would specialize in military government during hostilities as well as "post-hostility" rule. In May 1942 the army opened its School of Military Government and Administration at the University of Virginia, Charlottesville, where graduates specialized mostly in European problems of administration and governance. As the war spread east and troops came within distance of the Marshalls and Oceania, Charlottesville's people became desperate: "They didn't have one page of information on native government, nor did they have any idea where to begin. They want very, very badly some authentic information on just how colonies are governed," reported a Lieutenant Colonel Coulter of military intelligence.[2] Administrative problems of sanitation, control, resettlement, and food supply were "immediately pressing," and although the school's experts knew something about Europe, they knew next to nothing about Micronesia.

As the Pacific islands were expected to come under exclusive American rule, a second school, the School of Military Government and Administration of Columbia University, was founded in the summer of 1942 to administer them. The navy hoped that Columbia's graduates would be "the cream of the talent in civil affairs of the Western Pacific," and to confirm this elite status, high-ranking dignitaries made visits, while the school was given a nod from President Roosevelt and undersecretary of the navy James Forrestal.[3] Officially the military government school was under the umbrella of the Civil Affairs Committee, the overarching organization in charge of coordinating the supervision of all the various branches, planners, divisions, and schools dealing with occupation rule in Europe and the Pacific. But in the case of island governments, the navy very often devised policy on its own.

Quite self-consciously, the navy wanted both schools to be training grounds for an elite corps of American administrators abroad. Curriculum for a forty-eight-week course (later reduced to thirty-six) in military government and administration was drawn up, and an initial class of civilians—often university professors, engineers, or administrators—was "procured." The school's planners were confident that with the proper training and resources they would be able to breed or "engineer" new leaders. Great hopes rested on the schools of administration where, instead of ruminating about abstractions such as the proper role of rulers and governed, the navy put together a pragmatic modern experiment in governance.

Strikingly, to breed the new leaders, the navy used an organizational method very much like the behavioral science paradigm that cutting-edge social scientists had pioneered ten years earlier. It combined tactics from the fields of anthropology, psychology, sociology, psychoanalysis, and related areas to construct a science of social control for the management of human beings. The military government school offered combined training in anthropology, social psychology, sociology, and political science to its initial class of 150 naval officers. A navy press release announced it was "the most intensive course ever given in an American university" and that "educational history has been made" by the new methods. (Handling small arms and submachine guns, as well as supervision techniques, was another part of the curriculum; these had less precedent in the social sciences per se.) The navy set out to prepare graduates to govern "some remote island in the Pacific" and "to achieve order and harmony on the principles of honor and justice to the civilian population."[4] Graduates awaited billeting to the Pacific Forward Area.

The cross-disciplinary approach, new as it may have been to the armed forces, was not of course new to the social sciences. Various scientists had conducted laboratory experiments along cross-disciplinary lines for decades, out of which they had developed a paradigm for modifying and adjusting human behavior—especially its subjective or emotional aspects—in a particular environment. As we have seen, the turn-of-the-twentieth-century "engineering stance" of biologist Jacques Loeb toward life-forms in the laboratory was followed by the efforts of Watsonian behaviorists to transfer this approach to humans. In the 1920s the growing movement found a catalyst in Beardsley Ruml, who geared up a massive Rockefeller conduit for funding the social sciences. Spreading throughout the nation in different programs, the idea of human engineering through cultural design, anthropological data gathering, and "psychotechnology" came into its own.

At the forefront of this movement, the Yale Institute of Human Relations sponsored the most strenuous effort to merge psychoanalysis with behaviorism and to provide social scientists with access to the shadowy recesses of human interior life. A cadre of anthropologists contributed to the effort their databaselike "Bank of Knowledge." When war broke out, university scholars and scientists believed the time had come to put these experiments and information-gathering techniques into action. The files' anthropological authors, along with many other social science specialists attended the military government school at Columbia, whence newly trained academics and professionals soon carried out duties in administration, civil affairs, and military police in the occupied areas. While studying at the Columbia school, Murdock and his team were designated Research Unit No. 1 and produced a series of operational handbooks that helped naval strategists decide which islands to invade among a slew of hundreds: information from the files about harbors, beaches for landing, flat terrain for airfields, and of course Japanese entrenchments was supplied, collated, and produced. The unit was cited for "invaluable contributions to the limited knowledge of those areas"—that is, areas of the central Pacific that were, until recently, mysterious unknowns on the map.[5]

In January 1944 the first group of trainees in government affairs graduated, and as a navy press release announced, "to administer the territory the invading fleet carried a group of specially chosen naval officers trained for just such work." Soon Murdock, Ford, Whiting, and others joined Admiral Nimitz's fleet.

AMERICAN RULE OVER THESE ISLANDS, which were suddenly of great consequence, involved a question of style. Longtime British colonial civil servants came to lecture at the Columbia school and provided hard-won lessons from their years of service to the American administrators-to-be and officers-in-training. They began their lecture series by putting a diagram titled "The British Colonial Empire" on the blackboard, moving from H.M. the King at the top, to H.M. Government in the U.K., on down to the most menial colonial minion. But despite their rigid top-down recommendations, the civil servants in fact yielded to a more flexible and experimental approach. The anthropologist Arjun Appadurai offers an impressionistic sense, drawn from his own experience, of a parallel shift from British to American styles: "I did not know then," he writes of his teenage years growing up in Mumbai, "that I was drifting from one sort of postcolonial subjectivity (Anglophone diction, fantasies of debates in the Oxford Union, borrowed peeks at *Encounter*, a patrician interest in the humanities) to another: the harsher, sexier, more addictive New World of Humphrey Bogart reruns, Harold Robbins, *Time*, and social science, American-style."[6] That social science should be included with Bogey in comprising an American subjectivity is not the usual view of American attempts at nation-building. But it is perhaps more accurate than is generally acknowledged.

In the early months of 1944, as events unfolded on the Marshalls and Carolines, on Guam and Guadalcanal, and finally on Okinawa, finely shaded considerations of "dual mandates" and "mutual benefit" seemed like niceties. Battles raged, and island natives were caught between the outgoing Japanese and the incoming Americans. Kwajalein atoll in Micronesia was the site of the most concentrated bombing of the Pacific war; photographs from S.L.A. Marshall's rousing 1944 *Island Victory* showed a veritable wasteland studded with corpses, where those who hid in caves during the battle had to walk on top of the dead to get out. ("Never in the history of human conflict has so much been thrown by so many at so few" was a famous capsule of events there.) Air raids and blockades of remaining Japanese-held atolls Kusaie, Palau, and Ponape guaranteed a constant shortage of food even on bypassed islands. Napalm, in one of its first experimental uses, was dropped on several islands. The Marshalls were classified as vital intermediate objectives on the way to the conquest of Japan.

Winning the war was all. In the island governments military officials wrote internal memoranda admitting they paid inadequate attention to civil affairs. For the time being, America's general principles with respect to liberated areas were exigent. A draft memorandum from the head of the Columbia school put it this way: the first priority was "military considerations leading to total defeat of our enemies"; second was to carry out the "minimum" civilian relief and rehabilitation to prevent disease and unrest, which otherwise might cause civilians to interfere with military operations; and third, to assist in the restoration of the basic economy of the area, so as to lighten the burden on the military.[7] A civil affairs pamphlet titled "What Is Civil Affairs?" provided some background: the U.S. military had never previously had a civil affairs branch. Question: Why is it now necessary? Answer: "Firstly, as we are now waging total war, the whole of the civilian population have [sic] to be considered the same as any other military factor that is taken into account in making a plan."[8] That is, native civilians were one factor among many in a military calculus.

The status of the island inhabitants is suggested, for example, by their almost complete absence from the first several hundred pages of Dorothy Richard's encyclopedic *United States Naval Administration of the Trust Territory of the Pacific Islands*. Native welfare was often sacrificed to the forward motion of troops. Indeed, the official rules establishing military government stipulated that the health and well-being of natives be maintained for the express purpose of facilitating and accommodating the U.S. military presence.[9]

By 1944 the American fleet was island-hopping across the Pacific. Before any civil government could be established on an island (a task anthropologists would later help carry out), it had to be occupied militarily (a task with which anthropologists also assisted). Native-born Micronesians and Japanese settlers were "concentrated"—that is, placed in internment camps—"as a move to eliminate major points of friction and misunderstanding and to permit rapid construction of the base."[10] The occupying fleet recognized that removing the population to one island or a part of an island could be tricky, as the "recent agitations in this country relative to the movement of the Japanese on the West Coast" testified. Therefore officers were instructed to "use the utmost care in handling this problem," especially in regard to property Micronesians left behind in their homes.[11]

ALMOST TWO YEARS EARLIER, in March 1942, the War Relocation Authority forcibly corralled Japanese-American residents of the West Coast of the United States—specifically all those living in Military Area No. 1, the western parts of Washington, Oregon, and California and the southern part of Arizona—and gave authority over different camps to different government outposts. The Poston Relocation Center, near Parker, Arizona, was assigned to the Bureau of Indian Affairs, aka the "Indian Service," and was built from scratch on some mesquite- and creosote-filled Mojave reservation land. Like the five other sites, it was deliberately designated a "center" so as not to evoke concentration camps. The administration of these first-, second-, and third-generation Japanese-American "evacuees," as they were commonly called, became a concern for social scientists.

Five or six anthropologists and sociologists went to study the conditions among the 7,450 held at Poston. While high-up officials invoked the willing sacrifice that Japanese-Americans were making to their nation and praised their expected industriousness in making the desert bloom, the top anthropologist, Alexander Leighton, studied the challenges of administering a population that was unwilling to be there, much less be administered. Leighton argued in his *Governing of Men: General Principles and Recommendations Based on Experience at a Japanese Relocation Camp* (1946) that anthropologists could, by analytical methods, come to conclusions that were equally useful for science and for social control. Indeed, anthropologists working at the camps were provided the opportunity "for a degree of experimentation that is not possible elsewhere."[12]

Memos from the Columbia school describe how new military government officers used these detention experiences on American native ground as case studies. Anthropologist Leighton's book was written with the general format and tone of a "how to" guide for administrators, and it featured the frustration-aggression hypothesis as the basis of principles 1 through 7 for dealing with populations under stress: how frustration and aggression combined with fear may lead to crowd panic, scapegoating, pathological rumors, and rioting, and how these uprisings may be quelled. Leighton and his team trained civil affairs officers and administrators at Poston, and versed Japanese-American internees in social and anthropological analysis "so that they could be helpful in occupied areas of the Pacific, during or after the war."[13]

Likewise in Micronesia, soldiers and civil affairs officials needed to understand the workings of Japanese minds as well as native minds. On the Marshalls, one of the U.S. military's first conquests, soldiers gathered the people living there together and confined them to the upper end of Majuro Island, taking all their possessions, "including live pigs," into U.S. custody.[14] Interning and requisitioning were simple, for the islanders possessed very little; on nearby Bikini, for example, an inventory recorded "practically no living or eating equipment—one old sewing machine."[15] Complications were anticipated as the navy advanced toward Japan and islands such as Ryukyu and Okinawa, which were much more heavily settled with Japanese colonials, and so these early islands— relatively unassimilated within the Japanese Empire—were deemed test cases and pronounced successful. The military attributed this success in part to the thorough training of the administrating and police officers of the School of Military Government and Administration at Columbia.

Tragedy could not always be averted, however. On Saipan in the northern Marianas, the Americans had little opportunity to work with civilians at first. When the troops landed, native residents (Chamorros) fled to caves in the hills, convinced by the Japanese that Americans were barbarians. U.S. Marines urged them to surrender via public address systems on the cliffs and on an offshore ship, whereupon thousands threw themselves and their children onto the rocks and surf below, where they drowned.

POSTWAR POLICY toward the Japanese mandates, first occupied in 1945 and then brought under the administration of the United States as a "strategic" trusteeship in 1947, was initially all but nonexistent. Policy bigwigs were primarily worried about Japan and Europe, and these war-tossed bits of an obliterated empire had little in the way of potential economic exploitability. Strategically they had simply to be kept in possession. Into the lacuna stepped George P. Murdock, along with several other anthropologists, including Douglas Oliver, who headed the United States Commercial Company's economic survey of the islands. Murdock became the de facto author of U.S. policy for governing the 850-square-mile Trust Territory.

Unlike other experts, Murdock was not interested in economic profitability; he argued that the sugar industry was of value only insofar as it provided markets and jobs for natives. A 1944 Columbia School memorandum, almost certainly written by Murdock, reached the highest levels of policymakers and "established the postwar framework of colonial rule in Micronesia."[16] In it Murdock urged that a benevolent and paternalistic stance be taken toward Micronesia, one that allowed native structures of government and social organization to remain intact. In keeping with his prewar work at the institute, Murdock proposed using the islands as a place to conduct an "experiment" in creating order.[17] His plan was social science writ large.

In the summer of 1947 Murdock returned to the islands with a navy-sponsored phalanx of anthropologists and other scientists to conduct an experiment, hailed as the most complete study of a given place ever attempted. Through the Coordinated Investigation of Micronesian Anthropology program, a team of forty-one physical anthropologists, ethnographers, linguists, sociologists, and "human and economic geographers" from more than twenty institutions were dispatched over twelve island clusters in what its organizers called "the largest cooperative research enterprise in the history of anthropology."[18] Airlifted on navy planes and ferried on navy patrol boats, social scientists fanned out as if performing a military maneuver, to conduct field research in the U.S. Trust Territory and neighboring parts of Micronesia, including the Marshalls, Kapingamarangai, Kusaie, Ponape, Truk, Woleai, Palau and the surrounding southwestern islands, Yap, Ngulu, Ifalik, Guam, Saipan, and Rofa. Their aim was to cover the area completely "to avoid the confusion and cross-purposes of 'hit-or-miss' research," said director Murdock.[19] It was a fact-gathering mission on an unprecedented scale, and the results added to the files and thus to the sum total of knowledge. The goal was practical but not greedy, as Charles Dollard, at the Carnegie Corporation, reported of Murdock's views: "The fact that these areas are economically relatively worthless is, in his opinion, an asset since economic exploitation has been one of the main barriers to development of sound colonial policy."[20] Total war was to be followed by total anthropology.

The navy's hopes for this program provide a vivid sense of a new relationship emerging between military and scientific undertakings. A curious letter from Admiral Nimitz to the head of the National Research Council (through which Murdock was coordinating his program) explains why the navy chose to support Murdock's total anthropology scheme: "The Navy has always taken an active interest in all scientific research." Presumably as a subset of this long-standing curiosity, Nimitz went on to affirm the navy's interest in Murdock's plan: "the Navy not only has an interest in the program . . . for the results which may accrue to science itself but also because these results may have important aspects in plans involving military considerations." Thus for Nimitz the navy was and had been a friend of both pure science and practical plans. This assertion is not surprising in itself; what is interesting is that it carefully separates the two domains from each other even as they are lumped together under the navy's benign regard. The purer the science, the more readily military considerations could be pursued, the admiral seemed to say. Finally, Nimitz ended by "again assuring you of my personal interest in the general principle which underlies the initiation of your program."[21]

Another letter, this one from Rear Admiral P. F. Lee, struck a slightly different note, stressing the navy's "strong interest" in this "anthropological and related research" due to "a pressing need for the knowledge to be achieved, in connection with problems of Island Government." Again, he suggests that the fortunate convergence of the interests of science and government, now almost identical, fueled the research: "It is believed that the research interests of civilian scientists in the peoples of Micronesia coincide in large measure with the needs of the Government for specific information on these peoples."[22] Toward these coinciding ends, the navy gave $100,000 to fund the project. Supplying each of the researchers with a gallon of DDT, the navy ferried them to and from their assignments.

Even though both Murdock and the navy insisted that their objectives were "to a very large extent identical," this was not always the case. And even when their aims were identical, the degree of urgency with which they wished to pursue them was not, for, from the military's point of view, "the practical if not the scientific objective of the program renders imperative the early availability of the information secured."[23] Researchers were ordered to take notes in duplicate and forward the carbon copies from the field directly to the National Research Council, which would in turn pass anything urgent on to the navy, but they often missed their deadlines. Given access to officers' clubs and bachelor officers' messes, navy office space, food provisions, typewriters, and stationery, they found that these privileges sometimes interfered with their research. Adopting the point of view of "Milgovt" did not always sit well with the large fleet of anthropologists and "related researchers." Some were eager to make clear the terms under which they operated: the program, they informed the navy, "must make its position clear and emphasize the importance of the freedom of science." Science, however, implied not only freedom but duty, and another bulletin urged the participants, "All of you owe it to science" to make successful investigations, to avoid stirring up problems with navy or native personnel, and to make any criticisms of the former "with tact."[24] At times "science" was invoked to distance the project from the navy and at times to spare the navy unwanted criticism.

One of the forty-one researchers, David Schneider, made strenuous efforts to separate himself from the navy, but to his chagrin these efforts backfired and "only raised his colonial status higher," as Ira Bashkow explains in his account of Schneider's fieldwork under Murdock on the islands of Yap:

At a canoe-launching in December, wanting desperately "to be liked," Schneider "got tight" on toddy to avoid the appearance of "alignment" with the Navy's blue laws for natives. But if "refusing to drink" meant Navy, drinking also had implications: If Schneider broke laws willingly and with impunity, it could only mean that he was above them. To every subsequent ceremony, [the Yapese chief] Tannengin had Schneider bring two bottles of beer, which he would insist that they drink at a climactic moment. Wondering "why tannengin ostentatiously drinks beer with me," Schneider noted in February the old man's "advertising" that "I am of even higher rank than the navy."[25]

Each attempt to show the Yapese that he was independent from the navy and "on their level" led the Yapese to conclude that the anthropologist must in fact be *above* the navy in status. This conclusion was confirmed by the success Schneider had in persuading the navy to cancel road construction and withdraw certain personnel from Yap. The anthropologist soon found himself embroiled in local power struggles, unable to get "straight" answers from the Yapese, and finally, in disgust, he reverted to Murdock-style anthropology as the way out of his dilemma, painstakingly coding his voluminous data with the codes of the Yale file categories instead of devising his own interpretive scheme. On the practical side, the navy clearly got a wealth of information from the fieldwork Schneider and others did. Navy brass hardly ever left their colonial offices and certainly never went to live among the islanders; therefore much information on local succession, political battles, and Micronesian attitudes toward naval rule was passed on through the anthropologists.

Meanwhile another set of experiments unfolded, for the islands did not prove useless, contrary to what Eisenhower suggested in his "nothing but sandpits" comment. A- and H-bombs punctuated the American occupation, turning night into day and sand beaches into glass. In 1946 the first Bikini atomic test captured the world's attention: with its 140 planes, 200 warships, 200 goats, 200 pigs, 4,000 rats, and 42,000 servicemen, Operation Crossroads conducted an immense offshore experiment. The physician David Bradley worked for the Radiological Safety Unit at this "fascinating laboratory of Bikini," and after months of looking at the irreversible effects of radioactive fallout, he published his conclusions in the 1948 best seller *No Place to Hide*. To conceive the world in terms of a laboratory was eminently a social science point of view, but it was now also eminently a part of the United States' relationship to the world. The fact that two of Bikini's satellite islands had simply ceased to exist in the course of the experiment was seldom reported. Likewise it was hard to adequately assimilate the fact that the Bikini experiments marked the advent of a new atomic testing environment with which all humanity would need to reckon. (Anthropological dispatches from neighboring islands rarely mention the crescendo of test bombs.)

A few days before Christmas 1947, atomic bombs fell on Enewetak Atoll and three of its islands vaporized, going up in clouds. When, sometime later in November 1952, a ten-megaton H-bomb was dropped on Enewetak, top-secret conditions were in place. There would be no more public relations dramas or mass photo ops with native chiefs and assembled subjects. In 1954 an unexpected shift in winds and inadequate planning caused the Bravo disaster, which left ten centimeters of white radioactive powder to settle like snow on the island and people of Rongelap.

Note that each of these tests was "framed" as a laboratory experiment and that the island was the laboratory. (Sometimes, much to their misfortune, the islanders were de facto laboratory subjects, as an official report stated: "the habitation of these people on Rongelap Islands affords the opportunity for a most valuable ecological radiation study on human beings.")[26] Just as the Trinity Test in the New Mexican desert served as a laboratory for Hiroshima and Nagasaki—cities themselves chosen for the fact that they were "virgin targets" and so "almost like classroom experiments"[27]—so too by the postwar period field testing in laboratories ever farther afield became a national imperative.

Throughout this series of nuclear experiments, residents were displaced and suffered unfavorable conditions on the strange islands to which they were moved. Many islanders carried significant loads of nuclear contamination within their bodies, especially strontium-90 and cesium-137, concentrated particularly in children. These radioactive materials continued to enter the ecosystem and the bodies of the islanders' beloved delicacy, coconut crabs. Residents of the four "nuclear atolls" Rongelap, Bikini, Enewetak, and Utirik were warned not to eat more than one coconut a day, not to plant in certain places, and to avoid coconut crabs, but they still suffered miscarriages and deformed babies. Seventy-seven percent of Rongelapese under the age of ten at the time of the Bravo test developed tumors of the thyroid by 1979. Yet testing continued apace: in 1960 the Nike-Zeus missile system, in 1964 the Nike X, and in the late 1960s intercontinental ballistic missiles sailed across the "mid-corridor" of Kwajalein lagoon and landed in strategic impact zones directly on top of residential sites, now vacated. So it was that Kwajalein's twin island Ebeye became a colonial slum, its residents relocated from former test sites and mainly employed in the service industries that catered to American military personnel at the base. By 1978 reigning conditions there were summed up in a study titled *Ebeye, Marshall Islands: A Public Health Hazard*. Micronesians from different islands sought well over $5 billion in damages from the United States during the 1970s, and a nuclear-free political movement gained many supporters.

But all this lay in the future. From 1946 to 1951 the presence of Americans was still new, and its series of nuclear experiments was little protested. They were accompanied by the largest anthropological undertaking in history, and the question remains, what sort of science came out of this ambitious parallel experiment? Having spent several months on the island of Truk with four of his students studying native social organization, Murdock published several articles on this subject, including his rather mysterious 1948 "Symposium: Skeletal Muscle Relaxants—Anthropology in Micronesia." Another of Murdock's students, Ward Goodenough, wrote a classic study of social organization, *Property, Kin, and Community on Truk* (1951). He went on for several decades to employ research from the expedition to buttress influential theoretical statements, and to argue that the field of anthropology constitutes a sort of living bank of human information, the "database with which we work."[28] In addition, two documentary films emerged from the project, one of which, *Land of Mokil*, created controversy because it showed bare bodies and seemed to "editorialize" negative attitudes among the Mokilese toward increased trade with the United States. The filmmakers altered the film accordingly to show the Mokilese more fully clad and "wholeheartedly in favor of trade."[29] Physical anthropologists provided a somatology of the Ulithi atoll residents according to facial features—"Oceanic Negroid," "Negritoid," "Mongoloid," "Indonesian," and "Generalized Micronesian"—while their cultural anthropologist counterparts focused variously on myths, belt weaving, rituals, sibling rivalry, and minor oracles. Intensive fact-collection produced thirty-two full reports and more than one hundred articles and other publications. Much of the material was extruded and filed in Murdock's vast filing cabinet at Yale.

People's "inner states" and personality types seemed to transfix the researchers most. They brought a set of fourteen "specially adapted" Thematic Apperception Test cards with them, geared to help them evaluate life on tropical atolls. Each one featured a scene drawn by an artist in Chicago and was designed to so captivate the islanders that their collective descriptions of the contents would reveal their basic personality structures at work. One important study on the island of Truk used the cards to find out, in essence, who the Trukese were. Supervised by a psychologist, administered by an anthropologist, interpreted by a psychiatrist, and reinterpreted by the anthropologist, the whole appeared as the monograph *Truk: Man in Paradise*. Researchers gave a battery of psychological tests, including the TAT cards, to a little over one-fourth of the population. The results confirmed the basic ethnographic insight: the Trukese, while healthy in the main, lacked psychological depth and were mostly obsessed with finding food. A separate study on Ulithi also used the psychological test cards and yielded similar conclusions. Reasserted in several reports, the basic finding was that "Ulithians have, as their main goal, food and oral gratification."[30] In short, Ulithians were found to be more "concrete" and "enumerative" than the average American or European.

Concreteness meant that the native informant was likely to count the people on the card rather than delve into sexual or chaotic scenarios. A drawing of two men sitting side by side on a log—"which to many Americans might indicate . . . some sort of libidinal interest in each other"—prompted no Ulithians to think that a homosexual dalliance was about to occur, but rather stimulated thoughts about the weather.[31]

IN A REPORT for Murdock's Coordinated Investigation of Micronesian Anthropology titled "A Psychotic Personality in the South Seas," the psychoanalytically inclined anthropologist Melford Spiro studied a man named Tarev. When Spiro gave him a projective psychological test, the Stewart Emotional Response, Tarev performed so aberrantly that he was declared insane (a verdict with which his fellow islanders agreed, calling him a "fool fella"). His madness seemed to center on the Americans and on Spiro in particular: he made a pole to dance around to acquire their power (believing Americans "have something in their clothes"), followed the anthropologist everywhere, fell into trances during which he would deliver lines from American movies, and argued with voices in his head that told him the Americans were thieves ("they say that tree [pointing to it] does not belong to you, it belongs to us"). On the other hand, although he provided no evidence, Spiro believed Tarev's madness stemmed from the abuses of the Japanese. When the Micronesian expedition was about to depart, Tarev told the anthropologist, "If you leave here, I'll die," and he attempted to drown himself in shallow water. Spiro never returned and was unable to learn anything of Tarev's fate.[32]

The encounter between Spiro and Tarev raises the question of who the protagonists were in this new hybrid project of social science experiment and neoimperial government. The war had unsettled patterns of relating that had coalesced in different ways over the past three hundred years of Spanish, British, Australian, German, Japanese, and now American rule. (Different circumstances prevailed on each island—some had histories of intensive colonial contact, some had almost none.) At times newly arrived anthropologists were perceived as ghosts or spies coming from the land of the dead, but in general islander accounts of this period tell of Americans' (nonmilitary and military) generosity with their money and matériel, their willingness to share food in situations where previous colonial proconsuls had forbidden commensality, their easy rapport and unwillingness to stand on ceremony. There was in general far more reciprocity in relationships than before the war, and it is exactly the tension between such footing-in-friendship and the forever unbalanced situation that bears exploring.

What Admiral Nimitz or the navy, with its stated interest in "all science," made of these investigations is not recorded in detail. But one can conclude that three discernible things came out of this strange collision: U.S. policy for administering the islands and guidelines for anthropologists to aid in administration; a slew of anthropological publications and data for the files to process; and as Tarev's observations make clear, a new way of life on these "far-flung" islands. American rule was a sort of alchemy, characterized by what D. W. Brogan in his famous essay on the character of such rule called a "magical power of transformation."[33]

THESE EPISODES OFFER A VIEW of the strange concatenations that result when everything—the status of people, the status of land, and the status of reality itself—is up for grabs. They show the human engineering stance of the social sciences, and anthropology in particular, at work. This stance may be forsworn, but it is far from irrelevant. Its assumptions continue to operate in our own days of nation-building, nation-occupying, and regime-changing. To consider this brief history of mostly forgotten events from a faraway place is to reflect on the enduring urge to bring an American style of governance and American-centered goals to faraway places: often commendable, yet equally often ill-fated.

CHAPTER 10

The Impossible Experiment

AS THE LONGTIME HEAD of UCLA's prestigious Neuropsychiatric Institute, Louis Jolyon West was headquartered for many years in a suite of offices, with a trio of secretaries and a university car at the ready. Although "headquartering," with its command-central connotations, is not a term one can easily apply to many psychiatrists, it was apropos in the case of the larger-than-life "Jolly" West. Chief psychiatrist of the university's 250-bed hospital, he supervised a staff of more than fifty psychiatrists, while also managing at the peak of his long career to hold an additional seven hospital appointments, belong to about forty-five learned societies, and serve on nineteen committees, councils, and boards. Still, he was able to publish extensively, covering such diverse topics as hypnosis, brainwashing, hippies, racial violence, cults, mind-altering drugs, civil defense, homosexuality, elephants, scuba-diving, and the way of life particular to the Tarahumara Indians. "Reviews, critiques, papers, and monographs drop from him like hen's eggs," commented a journalist who interviewed Jolly during the Patty Hearst trial, in which Dr. West was the key expert witness for the defense.[1] Yet by the late 1980s, UCLA exerted pressure to force him to step down from the institute he had headed for twenty years. A senior colleague characterized him as "a high-level hack"; he had misused funds, he had made mistakes.[2] Most of all (this was the lurking context, mainly unspoken), it was no longer considered a good thing for a public university to employ someone who had worked intimately and enthusiastically with the Central Intelligence Agency.

West was one of many social scientists who began conducting research for the CIA in the mid-1950s during the heated years of the international "brainwashing" scandal. What distinguished West was the link he forged between the models of behaviorism and the particular interests and demands created by the cold war. His work flourished during a time when First World leaders were demanding new weapons to combat an enemy who was evidently capable of any violation of human dignity or code of decency—this much was clear from the willingness of the Soviets and Chinese to use brainwashing on their own subjects as well as on American captives. As the CIA's newly appointed director Allen Dulles commented in a 1953 address at Princeton University, Communist apparatchiks could render someone into a state of robotlike obedience in which the victim's brain "becomes a phonograph playing a disc put on its spindle by an outside genius over which it has no control."[3] Americans had failed so far to face up to these dire developments, Dulles continued, but now were compelled to respond in kind. There was an urgent need for American-designed brainwashing techniques that would do everything from "de-pattern" and rewire a single individual to mold and shape an entire people's attitudes and actions. It was not long before the Agency began its own program to explore "avenues to the control of human behavior," including "chemical and biological materials capable of producing human behavioral and physiological changes," as well as "radiology, electroshock, various fields of psychology, psychiatry, sociology and anthropology, graphology, harassment substances, and paramilitary devices and materials," according to a CIA catalog of its endeavors in this line.[4]

Over the course of his long career, Dr. West became an expert in some of the most dramatic transformations of which human beings were capable: he examined how, by means of high doses of LSD, intense psychological assaults, or environmental stresses such as long-term sleep deprivation, a subject could be deprived not only of personal autonomy but of such seeming fundamentals as identity and sense of self. In considering the most extreme altered states to which a human being could be subjected, West felt he had grasped the essential processes that produce a coherent sense of person and self—the "psychophysiology of conditioning." If one understood how people are broken, West reasoned, one could also understand how they are made and therefore how to change and control these processes. This research encouraged West to push toward a science of human engineering, a "truly unified behavioral science."[5]

Jolly West's career took shape at a confluence between science and government. In the mid-1950s the interdisciplinary social sciences—now grouped together as the behavioral sciences—were emerging from the academic sidelines into the national headlines. West was a catalyst for this movement, although he was far from the only one. During these years large portions of the academic behavioral sciences (sociology, anthropology, psychology, and psychiatry) came into contact with CIA operatives or were recipients of CIA money whether they knew it or not. In hearings held in 1945 and 1946 to plan the new intelligence agency, officials from the Office of Strategic Services identified these fields as "basic" social sciences that had contributed significantly to war activities and should be further encouraged.

It was natural enough that the interests of such scientists should flow into those of the Agency, for both groups were passionate about modeling systems of experimental control in the laboratory that could be used in real-world conditions. To "operationalize" what had for so long been a matter of hermetic lab work and elaborate experiments was a shared goal. Yet the waters where the two met did not always run smooth: there were obstacles, limits, opposing forces, and dangers. Careers, and in some cases even lives, were lost. And each individual scientist had his or her own protocol concerning what was proper and permissible in the service of country, humanity, and, not least, the furtherance of research aims and his or her career.

Among the most prominent advocates for a unified social science dedicated to mind and behavior control were three researchers, Jolly West, Harold George Wolff, and D. Ewen Cameron. These men, along with a cohort of like-minded experimentalists from neighboring fields, came to understand and later create altered states of mind and body. In doing so, they advanced some of the most ambitious plans for transforming human existence. All three doctors fairly dripped with the highest honors their respective fields could award, and although all worked for the CIA, it was never a seamless merger. Their stories show how a researcher's personal fascinations—what might be called today his "issues"—can combine with a sense of political exigency and historical moment to produce strange and ambivalent results. Called to service through science, West, Wolff, and Cameron found themselves facing the sometimes irresolvable, sometimes strangely energizing conflicts that arise with such calls.

JOLLY WEST'S LIFE revolved around the cliché that people who have the greatest insight into the American character are often those who are in some way outside it. Immigrants and their children are the most obvious examples. Eastern European Jewish immigrants' children such as Irving Berlin and Ralph Lauren created romanticized visions of an American way of life that they could see but not quite participate in. (Such visions are often preferred to the real thing, for Ralph Lauren's WASP clothing, as many have pointed out, looks a lot lovelier than the genuine article.) To be outside looking in has its advantages, and this is as true for scientists as it is for artists. It is perhaps the most true for *social* scientists. Jolly West was the child of Russian Jewish immigrants who initially settled in Brooklyn before packing up for Madison, Wisconsin. He grew up as the first male child, followed by two sisters, in a family living in Great Depression poverty. The military was his way out: after volunteering to fight at seventeen (serving as an infantryman from 1942 to 1946), he attended college and medical school at the University of Minnesota through the GI Bill. During the Korean War, he was chief air force psychiatrist at the Lackland base and also head of the department of psychiatry at the University of Oklahoma, having attained both posts before the age of thirty. At this point this promising young man who owed much of his promise to the government— "Without the government, I'd be just another lumberjack," he said once[6]—encountered a cause célèbre. An "extreme historical situation," as West's colleague Dr. Robert J. Lifton called it, was about to unfold: the cold war and the ultimately perverse interest it spawned in using mind control as a weapon to promote the cause of freedom.[7]

In the summer of 1953, with the Korean War armistice in place, twenty-one captured American soldiers announced to the world that they would not be coming home. They claimed they preferred to remain with their former enemy and raise families in Pyongyang or Beijing, for the U.S. political system and the hypocrisies it bred were repellent. The drama of "the twenty-one," many of them officers, created a public scandal as well as occasioning Virginia Pasley's book *Twenty-One Stayed*. Some months earlier, on February 21, 1952, twelve POW air force pilots confessed to flying "germ warfare" missions against North Korea. Their announcement, believed to be for the benefit of newsreel cameras, found a transfixed audience in America. Surely the twenty-one and the twelve had been brainwashed.

Receiving less news coverage was the fact that the remaining seven thousand or so men who had been captured by North Korea and held in camps behind the thirty-eighth parallel were also showing evidence of a problematic "attitude change." Behavioral scientists of all stripes were called in by the highest authorities in the Department of Defense, the State Department, and the military to assess, then address, what was shaping up to be a national emergency. In the immediate aftermath every single psychiatrist and psychologist in the service was diverted to work on the problem, assigned to task forces and research units. Preliminary studies conducted by social scientists on the returning men indicated that one in seven American POWs had been guilty of serious collaboration, such as giving tactical information; one in three had cooperated by signing propaganda leaflets or turning informer; and just about everyone, 95 percent, had in some measure failed to resist the enemy (for example, by complying with seemingly trivial demands or presenting themselves to interrogators as dyed-in-the-wool anticapitalists). More alarmingly, certain captive soldiers, upon their release, had been apprehended handing out pamphlets urging their comrades to desert; small newspapers in Indonesia and North Africa were featuring articles by American GIs extolling the collective way of life and Communist beneficence; and letters decrying capitalism's evils had made their way home. This, one *New Yorker* journalist declared, was "something new in history."[8]

Most of the prisoners had been captured toward the start of hostilities and held between twenty-four and thirty-seven months in prison camps run initially by the North Koreans and later the Communist Chinese. Upon their return through the exchange programs of Operation Little Switch and Operation Big Switch, they did not display the usual joy of the homeward-bound POW. An examining psychiatrist reported:

> When observed stepping down from the Chinese trucks at the American reception center, during the first moments of repatriation, most of the returning prisoners appeared to be a little confused, and surprisingly unenthusiastic about being back. During psychiatric interviews at Inchon just a few hours later, they presented striking consistencies in their clinical pictures. The average repatriate was dazed, lacked spontaneity, spoke in a dull, monotonous tone with markedly diminished affectivity. At the same time he was tense, restless, clearly suspicious of his new surroundings. He had difficulty dealing with his feelings, was particularly defensive in discussing his prison camp behavior and attitudes.[9]

Two weeks later, teams of psychological and sociological experts looked on as the USS *General Pope* docked in the San Francisco harbor. The experts recorded that released prisoners waited cattlelike on deck, unmoved even by the sight of their mothers' outstretched arms reaching toward the boat.

Whether or not the POWs' behavior warranted courts-martial or other legal action, great puzzlement reigned among America's military heads of staff: why had this happened when it had never happened before in any other war? Brainwashing, if that was what it was, had been successfully used on many captive Americans. The looming question to higher-ups in government and military was whether the Communists had developed techniques potent enough that no one could resist. A compliance rate that seemed rather close to total was troubling, and in the *Journal of Social Issues* two leading social scientists questioned "whether or not persons may at some time be helpless to control their behavior once they have fallen into 'enemy' hands."[10] Some critics were saying American soldiers were "soft," afflicted by affluence and mollycoddling; damning comparisons were made with the NATO Turkish POWs, who seemed to have borne up better under Korean control and shown greater resilience (dying at a far lesser rate, for example, from their injuries) than Americans while in captivity.

A vast deployment of "psy-warriors" was soon under way, as the army, navy, air force, Defense Department, and CIA assigned groups of experts to perform further evaluations on the released prisoners. Gathered in Washington, these scientists became acquainted with one another, and those with an interest in and willingness to participate in the CIA's behavioral control programs were identified. Almost all who were assigned to study the phenomenon of POW collaboration ended up in short order working for the CIA via one of its various "cut-outs," conduits, and false fronts, such as the Society for the Investigation of Human Ecology, the Geschickter Fund for Medical Research, and the Scientific Engineering Institute, or in one of its own laboratories.

Although the researchers did not agree on everything, they did broadly refute the popular belief that these men were cowards or had been subjected to the irresistible hocus-pocus of brainwashing. They found that nothing new had been practiced in the Communist camps, only good old-fashioned coercion, unstintingly applied. Opinion was divided, however, over exactly what sort of techniques had been used and exactly how successful they had been: Which soldiers had been most susceptible? Which were able to resist? Was a soldier bound to start spouting Communist formulas at home in Tulsa or Indianapolis? Some of the strongest of the resisters—the five percent of men who did not collaborate in any way, and who might otherwise have been considered the "heroes" of the camps—were often labeled by staff as antisocial to begin with, misfits known to rebel against authority in any situation. These soldiers were endlessly suspicious. They barely spoke a word during the group therapy sessions on board the returning ships and stonily ignored the psychological-test-givers.

WHATEVER HAD PRODUCED the high percentage of collaboration, the general thrust of the commissioned reports was that soldiers would have to be taught to fight it. This meant learning not magical counterspells but rather techniques to resist various forms of torture or "hard" conditions such as insufficient medical care, extreme heat or cold, sleeplessness, and the forced holding of uncomfortable positions for extended periods of time. Such pressures would have to be endured at least long enough—two or three days—that any vital tactical information would lapse in currency. Soldiers would need to be trained, in the popular language of the day, to act with honor; in the language of the social scientists, they needed to act in accord with knowledge of behavioral conditioning and how it works. In 1955, acknowledging this need, the armed forces adopted a new code of conduct that stipulated how a "U. S. Fighting Man" should act and what information—name, rank, serial number, date of birth—he could reveal under captivity. While the Chinese Communists who ran the camps violated the Geneva Convention (in effect since 1949), the existence and consequences of their actions couldn't be ignored. "It may sound trite, but if America is going to survive, Americans must learn to cope," said Hugh M. Milton II, assistant secretary of the army of manpower and reserve forces, in charge of one of the massive POW studies.[11]

In pursuit of this aim, West headed a group of researchers who were charged with training soldiers to understand and withstand behavioral conditioning. West's group studied the brainwashed airmen who had been returned in Operation Little Switch, America's first prisoner exchange with North Korea, which returned the most severely affected. As the lead researcher, West partnered with two young psychologists, I. E. Farber and Harry Harlow, to analyze the records of the men and to design future "survival training" courses. Of all the research groups formed by military branches and the CIA, West's group (started in March 1956) was notable for its thoroughgoing efforts to show how the conditioning experiments of the earlier part of the twentieth century were the key to understanding the phenomenon of brainwashing. Unlike other psychological experts, West stressed the connection to earlier advances in the social science of human engineering. Of course, journalists were keen to invoke the name of Pavlov—the scientist who in the popular mind represented dramatic success in behavior modification—and from the start they made incendiary statements that the soldiers had been conditioned just like the Russian scientist's drooling dogs, even as experts rebutted this "menticide" explanation as too simple. West's group, instead, devoted themselves to mapping out step by step the conditioning models used to change these men: Was it Pavlov, modified Pavlov, or Pavlov-plus-Watson-plus-Hull-plus-Dollard? How was it implemented? How far could it go? They made direct links with the earlier work of the Yale school of neobehaviorism, lavishly citing these scientists' work to explain the phenomenon of brainwashing.

West's group came up with the term "DDD" to explain what had happened to the Little Switch group of returnees. The first D was for *debility*: this was induced by semistarvation, fatigue, and disease. Often, as the prisoners explained in interviews, they were in chronic physical pain, their wounds left untreated. Poor health set the conditions for weakening the prisoner and making him even more amenable to his captors' demands. The second D was for *dependency*: this was produced by a prolonged deprivation of basic requirements such as food and sleep. The deprivation was interrupted by occasional, unpredictable brief respites, reminding the prisoner that the captor had the power to relieve his misery. The third D was for *dread*, produced by encouraging chronic fear—of death, pain, deformity, or permanent disability. Captors also hinted at the possibility of violence against a prisoner's family at home and further unnamed humiliations, and they even exploited the prisoner's inability to satisfy the demands of his interrogators as a further cause for anxiety. Such a degree of dread multiplied the physical pain a person experienced. In each case the factors of DDD sent the prisoner down a gradual path of weakening that in most instances led to total compliance.

According to West and his team, DDD worked because of "classical and instrumental conditioning."[12] West explained that this was best illuminated by learning theory, a type of behaviorism crafted in American laboratories just before the war. It was not pure Pavlov but was based on the work of researchers from Pavlov to Watson, Thorndike, Hull, Mowrer, Dollard, and Miller. Scientists at Yale's Institute of Human Relations had known for some time that one could induce "ultimate demoralization"—that is, behavior resembling a nervous breakdown—in a rat, a guinea pig, a cat, or a monkey by the skillful application and withdrawal of stress conditions. Within their laboratories, even punishment was preferable to the spiraling effects of an engineered environment.

The West team's argument was simple but powerful: the prison camps were a highly specialized environment, not very different from a laboratory in the sense that the prisoners were surrounded and contained, subject to manipulated stimuli. Different stimuli working in an almost totally controlled environment yielded specific behavioral patterns. At first the unfamiliar environment fostered apathy, a low-grade despair. In the camps when a person refused to eat or get up and look around and would not take any liquids, they called it "give-up-itis." At this point the prisoner could no longer see his way out of the prison camp world. When he became dysfunctional enough, breakdown or death ensued. The team explained:

> Whenever individuals show extremely selective responsiveness to only a few situational elements, or become generally unresponsive, there is a disruption of the orderliness, i.e., sequence and arrangement of experienced events, the process underlying time spanning and long-term perspective. By disorganizing the perception of those experiential continuities constituting the self-concept and impoverishing the basis for judging self-consistency, DDD affects one's habitual ways of looking at and dealing with oneself.

A translation might be: if you should find yourself in a Communist Chinese prison camp or a situation resembling it, you will get confused, lose clarity, and maybe fall apart. You may "regress" in language, thought, and even basic processes of knowing who you are. First you will run on autopilot; later you may be closer to a zombie.

The key to successfully brainwashing prisoners was not just causing pain but being able to provide relief. West et al. conducted extensive interviews that established prison camp patterns—patterns borne out by other researchers and by classic accounts of the Soviet and Chinese incarceration systems such as Arthur Koestler's *Darkness at Noon*, Alexander Solzhenitsyn's *The Gulag Archipelago*, and Robert J. Lifton's *Thought Reform and the Psychology of Totalism*. The procedure takes on a predictable dynamic. Say the prisoner is asked to confess his misdeeds and the evils of the capitalist system; he refuses. Day after day the pressure is upped, the situation is made intolerable; he is sick, he is not allowed to sleep, he begins to lose control of himself. But the minute he repents or makes even the tiniest concession, the pressure falls away. Sometimes there is tea and cakes, or the tender of a cigarette, or a handshake from his captors. This respite is followed by reassertion of tremendous pressure and stress. The moments of relief from pain, hunger, or isolation are experienced so strongly—even promises of relief cause what behaviorists call an anticipatory reaction—that they are powerful motivators. The subject takes on new behaviors and even thought patterns, patterns that are unlikely to die out until the prisoner's circumstances change radically. When pursued with sufficient zeal, such a program can bring about compliance over 99 percent of the time. "Under conditions of DDD, the possibility of resistance over a very long period may be vanishingly small," the authors wrote darkly.

The phenomenon West's team investigated caused them to address, in effect, many of the larger questions that vex philosophers and theologians: What is the will? What are its limits? Are we the "selves" or, alternatively, the "souls" we believe we are? Can that belief be changed, and if it can be changed 99 percent of the time, are we not for all intents and purposes lost? The West group's conclusions were both accurate and—when put to use in irresponsible ways, for example, as weapons against perceived or potential enemies of the state—frightening. For these psychological experts showed that what happens has little to do with any failure of "willpower"; it takes place almost automatically, a natural response of the human nervous system. It is not in any way conscious.

The team used their research to devise a training program—the Air Force Survival School—to help soldiers understand the mechanics of conditioning to some degree. The program was designed to dispel the "dread" component of DDD. For even if brainwashing in the pop-culture sense did not really exist—there were, as West and others insisted, no evil scientists from the Pavlov Institute hatching plots and polishing their scientific instruments—there certainly was such a thing as "forceful indoctrination," which through the step-by-step application of DDD produced remarkable transformations.[13] The survival school put trainees in stressful situations (dropped them at night in the desert with no food, captured them, and held them for two days in bewildering circumstances, during which time they endured treatment at the official border between "hardship" and "torture") to simulate what might happen in war. West's group also devised a scale of compliance to measure the degree to which a prisoner had cooperated with the enemy.

Later, with CIA involvement, DDD research was used in "setting conditions" for interrogation and for experiments in mind control. Meanwhile West was increasingly convinced that if behavioral conditioning techniques could be used to destroy, they were also the essence of healing. With this in mind, he redesigned the curriculum of the University of Oklahoma School of Medicine, which he headed until 1969, to include an intensive course of study of the basics of behavioral sciences—psychophysiology, learning, experimental sociology, anthropology, animal behavior, and human ecology. He envisioned a future science of human biosocial engineering that would work prophylactically and preemptively. Potential criminals, juvenile delinquents, schizophrenics, and drug addicts would be monitored through remotely sensed electrodes implanted in their brains. "The prediction of dangerousness"—the likelihood that a person would commit a violent crime in the future—"will be increasingly refined and quantified, although never, of course, perfected," West wrote.[14]

TO SCIENTISTS LIKE WEST, brainwashing was not just alarming; it was *interesting*. To be brainwashed was in a sense to die, for the "self" as one had known it ceased to function. And as West's group pointed out, DDD was analogous to other extreme states of consciousness and bore an "interesting resemblance" to postlobotomy syndrome—that is, the disorientation and identity confusion noted after the frontal lobe of one's brain was pierced with a surgical incisor. It was also akin to certain drug-induced states of mind, such as those resulting from LSD and sodium amytal. Of further interest was brainwashing's resemblance to schizophrenia and hypnosis. All these conditions—whether induced by chemicals, surgery, madness, trance, or coercion—appeared similar. The self was gone and yet the person persisted; in most cases the old personality eventually rebounded. If for scientists this phenomenon tended to raise existential questions (about the ontological status of the individual, for example), for the CIA it raised operational questions about how its effect could be minimized in some cases or produced in others.

The insights of West's group and the other "brainwashing" specialists redirected agendas at the CIA and in branches of the military. Human material was changeable, and men gone off to fight were susceptible. These realizations were frightening but also bracing. If indeed the world was rife with threats to the inner as much as the outer man, then experts in these realms were needed more than ever. Many good and well-meaning professors—self-described or de facto human engineers—participated in the CIA's programs to bring about slow or rapid change in the minds and behavior of people. In this way human engineering came to fruition.

In the welter of panic, excitement, and theorizing caused by the brainwashing scandal, the CIA authorized secret directives to explore behavioral engineering and mind control. In 1952 the long-standing project ARTICHOKE was redirected to examine the knotty question of how to get information from a person against his or her will. In 1953 the newly formed project MK-ULTRA began a more exploratory, anything-goes program. Eventually comprising a total of 149 sub-projects, MK-ULTRA supported research on the workings of certain stratospheric drugs; the hypnosis of secret agents, programming them to carry out missions unaware; the possibility of mind-control machines; the possibility of carrying out mass brainwashing, coercion, or subtle attitude adjustment and behavioral modification; the use of electroshock, intensive drugging, and lobotomy to control or "drive" an individual; the effects of extended sensory deprivation on one's state of mind; and the possibility that any of the above might be an effective interrogation tool or a way of making someone forget having been interrogated in the first place. In short, if you happened during these years to be a scientist examining the way controlled environments affect the mind's circuitry or the body's behavioral patterns, your work was of interest to the Agency.

On the CIA's payroll, West became a major investigator of the effects of LSD on personal coherence. For the researcher interested in the self and identity, no better subject afforded itself than the uniquely powerful LSD, which had been synthesized in the laboratories of the Swiss pharmaceutical company Sandoz in 1943 and which was causing an uproar among those select people (mainly well-connected writers and world-traveling spies) who knew of its existence. It was more potent, ounce for ounce, than almost any other substance. A tiny dot could do away with one's moorings in reality itself. Not surprisingly, the CIA was making a great effort to find out what LSD's properties and its potential uses might be.

Such studies were farmed out, with Sandoz-made stock going to Robert Hyde's group at Boston Psychopathic, to Harold Abramson at Mount Sinai Hospital and Columbia University in New York, to Carl Pfeiffer at the University of Illinois Medical School, to Harris Isbell at the NIMH-sponsored Addiction Research Center in Lexington, Kentucky, to Harold Hodge's group at the University of Rochester, and to Louis Jolyon West at Oklahoma and later UCLA, all of whom made use of CIA funds channeled through the Macy Foundation or Geschickter Fund for Medical Research, or sometimes through the navy or the National Institute of Mental Health. Hundreds of studies tested the drug's effects, altering factors within either the environment or the organism. Almost every type of animal was brought into the lab for testing, but the results were not spectacular except among certain spiders who were induced to spin the most obsessively perfect webs. Carp under its influence changed from being bottom dwellers to sur-face swimmers. Lab rats showed a tendency to agitation, other animals a lowered threshold for sensation. In humans the results were more complicated and actually quite hard to characterize: LSD affected everything in the organism—breath and heartbeat, organs and cells, brain and body—in a complex "maze of interactivity," as one re-searcher summarized it.[15]

While still at the University of Oklahoma, West investigated LSD's "psychotomimetic" properties—that is, the extent to which it made a person temporarily crazy. He capped off years of such investigations with what promised to be a bold coup on the animal-research side. Re-searchers had hitherto been content to administer doses of acid to na-ture's smaller representatives; in 1962 West delivered the largest dose of LSD in history via rifle into the hindquarters of a male Asiatic elephant named Tusko, hoping to find the secrets of the periodic "musth" mad-ness to which male elephants are inclined. But the animal underwent a prolonged and unexpected death—its larynx swelled shut, and it suffo-cated. Some years later West conceded, "We must have miscalculated the dose."[16] Undaunted, he continued studying LSD's effects on college students and, starting in 1967, on hippies, who did not need encour-agement to take the drug. "To study them in their natural habitat," West and his associates explained, "we established an apartment or 'pad' as a laboratory in the Haight-Ashbury district of San Fran-cisco."[17]

Throughout the heyday of these clandestine programs, researchers had free rein to perform what the historian Jill Morawski has called "impossible experiments," and what a 1964 CIA memorandum grouped under the heading "Sensitive Research Programs." Three areas of research in particular, writes Morawski, had been subject to unspoken "social rules" and taboos in American society—hallucinatory drugs, psychosurgery, and prolonged sensory deprivation—all of which became prominent topics among the CIA social scientists and researchers.[18] Their engagement in such topics was sometimes indirect and delicately oblique (for example, by choosing to study animal subjects instead of humans, and by remaining in the dark or half-dark about the CIA's involvement in their research) and sometimes direct, with a disregard for research subjects that the word *cavalier* only begins to characterize.

UNLIKE JOLLY WEST, whose dramatic approach to research complemented the swashbuckling style favored by CIA agents during the high-cold-war era, Harold George Wolff was extremely focused and an unlikely candidate to conduct secret mind-control experiments. In 1953, when the brainwashing scandal broke, he was a well-respected physician and a world-famous neurologist. He quite literally wrote the book on pain—his *Pain* (1952) set a new standard for the study of how pain affects the human body and how it can be managed. But his real specialty was migraines. Even today his experiments on migraines are hailed for their creativity. He experimented on different subjects, including volunteers, himself, his lab assistants, and medical residents, as well as a flying trapeze artist from the circus who suffered from headaches but claimed that standing on her head made them go away. In 1947 he invented a machine for headstanding and ventured to twirl headache sufferers on a human centrifuge, both of which seemed to alleviate the headache for a while. A diminutive man with a nondiminutive personality, Dr. Wolff held a secure professorship at Cornell University Medical College in New York City for thirty years that kept him engaged and passionate about his topic of expertise. "Never a day without an experiment" was the motto emblazoned across his laboratory's bulletin board.

Born in 1898 to a Lutheran mother and a Catholic father, Wolff grew up poor but respectable and, after considering but rejecting the priesthood, took up the modest profession of dehydrating fish. During the 1920s, however, he switched to the study of medicine and was admitted to Harvard; a year studying psychiatry at Johns Hopkins and another year studying neurology capped off his graduate studies in the States. Travels followed to the laboratories of the European "greats"— Otto Loewi in Austria and Ivan Pavlov in Russia, as well as Sir Thomas Lewis in London. On his return in 1932, Wolff secured a position as head of Cornell's neurology department, which he held until his death in 1962.

Throughout his life Wolff embraced a puritan work ethic in laboratory terms, setting a disciplined pace and schedule from which he never deviated—playing squash every day, doing experiments every day, and performing other rigorous work. He managed to publish more than five hundred articles and thirteen monographs, of which, as a colleague remarked, "none are superficial."[19] He was editor-in-chief of his field's major scientific journal, the *Archives of Neurology and Psychiatry*, was on the staff of an additional half-dozen journals, served on twelve national medical committees, was active in twenty-seven medical societies, consulted at three metropolitan hospitals in addition to his post at Cornell, and kept up a continent-hopping lecture schedule—topping all this activity off in 1960 with the presidency of the American Neurological Association. A giant in his field, he was also a man-for-all-disciplines, combining social anthropology, clinical psychology, medicine, and psychiatry.

In the mid-1950s Dr. Wolff also directed one of the most ambitious arms of the CIA's behavior-control and mind-control programs. By then, he was boasting that he could provide the keys to "how a man can be made to think, feel, and behave according to the wishes of other men, and, conversely, how a man can avoid being influenced in this manner."[20] Dizzy with the possibilities, Wolff, trained in Pavlov's behaviorism and its American successors, gained unparalleled research opportunities scarcely dreamed of, much less acted upon, in more temperate times.

These opportunities first arose in late 1953 when Allen Dulles's twenty-three-year-old son sustained a serious head wound while fighting in Korea and Dr. Wolff's expertise was sought. Difficulties in the patient's convalescence and recovery made the relationship between father and doctor unusually close, and the two became good friends. With the trust they had established and the brainwashing furor in full flower, Dulles asked Wolff to conduct for his Agency a comprehensive study of Soviet and Chinese methods of brainwashing—known in more pedestrian language as "coercive persuasion" or "forceful indoctrination." Dulles gave Wolff full access to all the information the CIA had on this topic, including classified files and access to former KGB and Chinese Communist interrogators as well as the names of former prisoners. A group of twenty psychological warfare experts helped with the study as well, while a White House aide assured Cornell's administration that Wolff's work was as weighty and hush-hush as he had imputed. "It was done with great secrecy," recalled Wolff's partner in the study, Dr. Lawrence Hinkle, some years later. "We went through a great deal of hoop-de-do and signed secrecy agreements, which everyone took very seriously."[21] The resulting study became a classic in the field. When an unclassified version called "Communist Interrogation and Indoctrination of 'Enemies of State' " was published in 1956, it was the most influential of all the studies commissioned by the CIA and the military.

The published version strongly suggests that Wolff and Hinkle were given license to conduct their own brainwashing experiments. The set-up was along the lines of the old rat-in-a-maze but used human beings. A person was placed in an untenable situation—a "situation of frustration," the authors called it. Details are few, but any effort on the part of the test subject to escape or reduce discomfort was clearly fruitless. At first the subject diligently attempted to find a way of relieving the pressures placed upon him. However, "if one arranges the experimental situation so that the man cannot find a satisfactory solution by his exploratory activities," the authors wrote, a change could be observed:

> his next reaction is an increasing and random exploration, with a general increase of motor activity and an overflow of this activity into other behav-

ior, of a nonpurposive nature. He appears to "become excited" and shows evidence of anxiety, hyperactivity, and sometimes panic. If the pressures of the experimental situation are continued, the hyperactivity of the subject will gradually subside, with the exception of isolated repetitive acts. He may settle upon one form of response, which he repeats endlessly and automatically, even though this endlessly repeated action can never produce a solution. If the pressures are continued long enough, his ultimate response is one of total inactivity. He becomes first exasperated, and finally dejected and dependent upon anyone who offers to help him. He becomes unusually receptive to approval or human support.[22]

The subject becomes what psychiatrists call "emotionally bankrupt." He is ready to be transformed, ready to "accept suggestions which he previously would have rejected."

After bringing about this level of demoralization in their subjects, the Wolff-Hinkle experiments—at least those aspects that appear in the unclassified version of the report—broke off. Wolff and Hinkle conducted interviews with practiced interrogators to determine the remaining steps toward full compliance. Pushed past random exploration, past hyperactivity, past automatic repetition, past despair and utter inertness, the subject reached a point at which he became desperate for some human contact. Typically he started a deluge of talk, and a warm and dependent relationship with his captors developed. Rewards were bestowed, and an intense desire to please was the response. Confessions were signed, self-criticism more eagerly engaged in. If the relationship developed further, through positive reinforcement, a close bond resulted, one that resembled, in many respects, love. Some prisoners at this point went obligingly to their deaths—here was the mysterious aspect of the show trial, in which loyal revolutionaries confessed their sins against the people and welcomed their own execution. In cases where a reformed penitent rather than a dead puppet was the desired outcome, a type of religious conversion might occur, and some people experienced an unwonted exaltation.

While carrying out this research, Wolff seems to have believed in its importance for an enhanced science of human behavior. The tools of the interrogator, he felt, could be used in a strategic as well as a thera-peutic manner. On the one hand, headache sufferers could be cured by being forced—within a laboratory environment of extreme manipula-tions and controlled stimuli—to change their habitual thought pat-terns. On the other hand, the nation itself could benefit from the skilled application of such techniques to enemy agents, Communists, and subversives. Wolff proposed setting up a closet-CIA organization called the Human Ecology Society that would be authorized to do se-cret untoward experiments, with himself as president (a position he assumed in 1955). He created a new field he called "human ecology" that combined the disciplines of sociology, medicine, psychiatry, and anthropology. Since these fields had different views of the social and individual processes by which human beings become what they are, their combined knowledge, Wolff promised, would give the CIA the tools to control human beings. As Dulles had said of Communist brainwashing capabilities in his 1953 Princeton address, "We in the West are somewhat handicapped in getting all the details. There are few survivors, and we have no human guinea pigs to try these extraor-dinary techniques."[23] Soon that situation was rectified. Social scientists like Wolff and West—and scores more via the Human Ecology Society (later the Society for the Investigation of Human Ecology), MK-ULTRA, and other directives—would provide guinea pigs, human and otherwise.

HOW DID DR. WOLFF FIND THE ENTHUSIASM, much less the time, to conduct such investigations, which did not appear to have much to do with headaches? An answer involves considering exactly what he thought about headaches. In fact, Wolff's experience (since childhood) and theory (since 1933) of the migraine was very broad and almost philosophical: he felt that such headaches resulted from an imbalance between the individual and the world, the two being out of joint, cut off. People with migraines tend to shut off or strive to control their environment because they are so driven and high-achieving, so "good" in the terms society demands and expects. When this shut-off becomes too extreme, the head in effect answers back. First the vascular system spasms, producing the "aura" (blind spots or strange lights that indicate the onset of migraine). Next the vascular system dilates; arteries deliver more blood to the brain; other tissues respond as well. A headache results. It is not only a physical event but a total event. Pain sensitivity thresholds are remarkably lowered. Afterward equilibrium is achieved, if only temporarily, and the patient is in good spirits, even ebullient. Wolff theorized that a migraine headache was an adaptive response, however misdirected—the brain thought it was being attacked and tried to protect itself. Wolff's theory—the "neurogenic concept of vascular headache" (that migraines are caused by unstable cranial vasomotor functions)—was fully elaborated during the 1950s; the fact that the CIA was funding part of the necessary research was not mentioned.[24]

Wolff's human ecology experiments began when a certain number of his patients volunteered for a new experimental approach. This first group was composed of people with migraines (caused, according to his theory, by a disharmony between individual and environment), whom he accordingly placed in a disorienting state in order to reprogram them and create new behavior patterns. Wolff likened his role as doctor to that of a Communist interrogator and explained that, like prisoners, his patients were reluctant to change their ingrained habits. In an effort to make his patients more receptive to his suggestions, Wolff tried sensory deprivation chambers where he kept headache sufferers locked in until they showed "an increased desire to talk and to escape from the procedure."[25] Indications from CIA records are that Wolff's patients emerged terrified and mentally disorganized but lacking, also, their original headaches.

In a more ambitious experiment in 1954, Dr. Wolff gathered together a group of one hundred Chinese refugees to test more broadly his understanding of individual-environment interactions—to wit, that the two are in constant interplay, and that altering the environment in specific ways can produce specific changes in the individual psyche or sense of self. This experiment was also intended to reprogram the refugees and turn at least some of them into secret agents who could infiltrate Communist China. The subjects, who were mostly middle-class escapees from the Maoist regime, reported to the hospital and were paid twenty-five dollars a day to take part in interviews and be subjected to stress-producing situations. In fact, Wolff's grander aim was to train them to withstand future attempts to condition them should they be captured: they would be preprogrammed—"preconditioned," that is, preemptively brainwashed in perpetuity, or so Wolff claimed. Anthropologists, psychologists, and psychiatrists signed on to the project, which was given high priority by the CIA's internal experts on behavior and mind control. (Many participating researchers had no idea they were working for the CIA, were not permitted access to certain classified parts of the experiments, and thought they were conducting basic personality and culture research, which normally involved exploring how childrearing and early conditioning affects the formation of adult personality. Lower-level researchers were lied to by Wolff, who denied any CIA connections.) Initial results were inconclusive, and no anti-Communist agents appear to have come out of the outlandish program. More significant was the human ecology laboratories' matter-of-fact view that brainwashed secret agents were a realistic and most desirable goal—the only way to trust a foreign operative, in effect, was to possess utter control over his or her mental processes. To Wolff's disappointment, the study was interrupted by an administrative shake-up and was never completed.

Although Wolff in the last half of the 1950s continued to receive almost $300,000 from the Agency for his research on the brain and central nervous system and continued to serve on Human Ecology's board of directors, he was to some extent sidelined by an opposing and increasingly dominant camp within the CIA's Technical Services division. It was to this camp, in fact, that Human Ecology itself was turned over. Confidence in Wolff seems to have faltered among his scientific peers within the Agency, if not among the high-up bureaucrats who continued to be his friends and allies. The change in Human Ecology may also have resulted from infighting among CIA bureaus concerning some of the farther-out experiments being run within the Agency by its own scientists and operatives or outside the Agency by its hand-picked researchers. Areas of concern included a spate of in-house acid-dropping at the Langley headquarters and the subsequent death of the biochemist Frank Olsen, who had unwittingly been sent on a "trip" and fell into suicidal depression. Some expressed the view in private memos that the MK-ULTRA staff was becoming unhinged. A more cautious regimen was imposed. Meanwhile, in 1956 Human Ecology was removed from any affiliation with Cornell University (Wolff's home institution) and was set up to function as a more "legitimate" organization from then on, devoted to funding cutting-edge behavioral research and inquiring into the results but not directly interfering in the conduct of experiments. It was run by a more impartial board and concerned itself not simply with Wolff's research but with anything that could be harnessed for MK-ULTRA, ARTICHOKE, or other programs.

The CIA chemist who headed Technical Services, Sidney Gottlieb, took a more systematic and sophisticated approach (compared to Wolff's wild promises of achieving total control over a human being), with a more methodically pursued goal of exerting influence over or manipulating subjects. He believed small steps toward small changes in personal behavior tested in and out of the laboratory would add up to major results. Allen Dulles continued to come to Human Ecology meetings, and many, many scientists continued to examine the workings of behavioral conditioning on mental states. The myriad projects Human Ecology funded included everything from the compendious volume *The Manipulation of Human Behavior*, edited by brainwashing experts Albert Biderman and Herbert Zimmer, to the renegade work of the sociologist Erving Goffman who, in writing his brilliant *The Presentation of Self in Everyday Life*, explored the quixotic nature of human identity on the CIA's dime. (Goffman, like other left-leaning social scientists, did not know the ultimate source of the "publishing grant" he received.) The Macy conferences once a year brought together luminaries such as anthropologist Margaret Mead and child psychologist Jean Piaget for get-togethers at the Princeton Inn on appealing topics such as "Problems of Consciousness"—with undercover CIA reps and military contractors in attendance soaking up what they could. Research on hypnosis, interrogation techniques, isolation chambers, and LSD were the order of the day. Many leftists and liberals who otherwise had difficulty securing financial support (their research branded by most funding institutions as too "radical") kept going by means of Human Ecology money, including sociologist Jay Shulman, anthropologist Edward T. Hall, and psychologist Carl Rogers. Most researchers who received the smaller Human Ecology grants had very little idea where their funding came from, although it would have been easy to suspect—Shulman, who received a Human Ecology grant giving the CIA access to his leftist Hungarian research subjects, commented in retrospect: "My view is that social scientists have a deep personal responsibility for questioning the sources of funding; and the fact that I didn't do it at the time was simply, in my judgment, an indication of my own naïveté and political innocence, in spite of my [leftist] ideological bent."[26] Or, as others later remarked, they should have known and might have known but didn't.

The extraordinary latitude that the CIA gave Wolff to try out his ideas and the confidence it displayed in assigning him sensitive projects is remarkable but easy to understand in light of his unimpeachable reputation and his hard-to-touch credentials, not to mention his circle of friends. "From the Agency side, I don't know anyone who wasn't scared of him," a longtime CIA associate recalled. "I never knew him to chew anyone out. He didn't have to. We were damned respectful. He moved in high places."[27] Many graduate students and lab technicians treated him with tremendous deference, even revering him. Dr. Wolff was also something of a public figure, publishing popular articles in *The Saturday Review* and elsewhere on how a person's own stimulus-response patterns can create ungovernable fear or, then again, if properly controlled, the hope of changing oneself for the better.

FROM THE MID-1950S to the early 1960s, the CIA continued to fund experiments at the very edge of the ethically possible, and experiments that would not be acknowledged, much less defended in public, were funded in secret. Ice-pick lobotomies were performed by a surgeon in Washington, D.C., on a few hapless interrogation subjects, according to a 1952 internal CIA memo, and resulted in "nervous confusional and amnesia effects"—with the advantage of not leaving a scar.[28] In the realm of possibility were not only the standard trio of drugs, psychosurgery, and hypnosis but also any number of potential methods for altering behavior and mental constitution. Scientists went about using the old behavioral engineering paradigm—control the environment and you will control the organism within, the injunction to "study man in relation to his total environment"—but took it to extremes. They carried out long and arduous tests that involved breaking down a personality, introducing new programming, and bringing the subject back to functioning once again. The primary work was done at satellite laboratories.

At the root of the experimental urge was a commonly held fantasy, perceived as an impending reality, of being able to achieve absolute control over someone. Experiments along these lines were performed by Dr. Ewen Cameron at a special facility of McGill University, the Allan Memorial Institute, on the outskirts of Montreal. There, in the mid-to-late 1950s, attempts to rewire, reprogram, and remake human subjects went well into the murky territory where science meets sadism. Dr. Cameron, the "outside genius" leading these experiments, was at various times during his career elected president of the Canadian, American, and worldwide psychiatric associations. Perhaps the most remarkable of his honors and professional activities—in light of what followed—was his service as a member of the Nuremberg tribunal that prosecuted Nazi doctors.

Founded in 1943 with generous amounts of Rockefeller Foundation money, the Allan Institute was intended to lead the field of psychiatry in humane methods and take a refreshing sweep-out-the-closets approach to mental disease. It originally seemed to be a progressive and exciting mental institution geared to meeting the demands of an emerging postwar order. As its head, Ewen Cameron concerned himself with the battle stresses of soldiers returning from World War II and the ways their anxieties could be eased. Known for his generally decent treatment of patients, Cameron announced a progressive no-locked-doors policy even in the most severe wards, so that patients would never feel "locked away" and would always remember that they were part of society. Even during the institute's heady early years, however, there were troubling signs that Cameron's personal style and ambitions were less beneficent than his policies appeared: foundation officers were disturbed by the doctor's "failure to establish warm personal relations" with his peers and by a personal style that bespoke "a need for power which he nourishes by maintaining an extraordinary aloofness."[29] This might have constituted a warning that studies in which Dr. Cameron cultivated an extreme degree of power over his patients could go awry. Still, notwithstanding his cold bedside manner, Dr. Cameron was widely considered the ablest psychiatrist in the land, and patients were referred to him from all over Canada. A disaffected housewife, a rebellious youth, a struggling starlet, and the wife of a Canadian member of parliament were a few of more than a hundred patients who, shortly and without their consent, became experimental subjects.

Cameron took what was known about brainwashing—that one could bring about remarkable changes in attitude, ideology, and even a person's plain old habits through the imposition of highly controlled conditions—and pushed it to its logical end. As he claimed in scholarly articles, he treated his patients as the equivalent of prisoners who, for their own good, had to be broken down into infantlike nonpersons, so that they could be built up again. He likened his treatment method to the way an interrogator brainwashes an individual under continuous interrogation in wartime. Cameron's research was of great interest to the CIA, for it asked the persistent question, Is it possible to wipe a person clean and start again?

In the mid-1950s Cameron began to explore what he called "psychic driving." Soon, with an air of understated but confident discovery, he touted this technique in the major peer-reviewed journal of his field as a "gateway through which we might pass to a new field of psychotherapeutic methods."[30] The technique was to play for the patient over and over a "loop" of one of her own statements from therapy—a key statement on a major topic between five and seven seconds long with the tape running at the standard rate of seven and a half feet per second, as Cameron specified with the exactitude of one for whom the magnetic tape recorder was a great boon. After thirty minutes of playing the tape or "driving," a marked "penetration" was generally achieved, making hitherto inaccessible psychological material accessible to the therapist. In other words, the patient experienced an escalating state of distress that often caused her to reveal past experiences or disturbing events long buried. As patients frequently were disinclined to listen to the "driven" material, Dr. Cameron deliberately tried to thwart this avoidance by "use of pillow and ceiling microphones," the substitution of different voices for delivering the message (perhaps the mother, perhaps a peer), the administration of drugs such as sodium amytal, Desoxyn, and LSD-25 (the latter recommended for its ability to "disorganize" thought patterns), and the imposition of prolonged sleep. At other times patients were isolated in a sensory deprivation chamber: kept in a dark room, his eyes covered with goggles, his auditory intake reduced, and "prevent[ed] . . . from touching his body—thus interfering with his self image," a patient found himself in circumstances where, finally and most alarmingly, "attempts were made to cut down on his expressive output." In short, he was not even allowed to scream. All these factors could be varied or combined with extended periods of "psychic driving"—up to ten or twenty hours per day for ten or fifteen days at a stretch. Cameron reported that once the patient's resistance had been conquered, the result was therapeutic.

At the height of the CIA's mind-control efforts, Cameron's "psychic driving" article in the *American Journal of Psychiatry* was like a banner waving in the wind: to CIA scientists, Cameron was clearly someone to watch and support. How could they deliver that support? He could not be appealed to as a patriot, for he was neither an American nor a Canadian citizen. (He was Scottish, had his home in Lake Placid, New York, and worked on the other side of the border.) At any rate, a patriotic appeal was not really needed in this case: unprompted, Cameron was doing the very experiments on extreme conditioning that the CIA wanted to be done, post-Korea. A few months after the publication of

his "psychic driving" prolegomenon, Human Ecology in its latest incarnation granted him a fairly large sum that, in combination with the
other, larger amounts he was receiving, made him flush with funds.[31]

From that point on, his efforts intensified. (Why this happened is a
mystery. Perhaps the incoming American support encouraged him to
attempt to produce more dramatic results. Perhaps the experiments
produced their own momentum—the further he ventured into questionable methods, the more they appeared to be possible.) In 1958
and 1959 fifty-three of Dr. Cameron's patients, without their consent,
were in fact wiped clear of many of their memories, personal habits,
prejudices and neuroses, and even self-knowledge—after which they
attempted to resume their lives. Most had been diagnosed as schizophrenics (although this diagnosis has since been called into question
in a number of the cases).

The treatment was first to knock them out with "sleep" drugs for
around two months via a cocktail of Thorazine, Nembutal, Seconal,
Veronal, and Phenergan, intermittently "depatterning" them with electroshock and frequent doses of LSD to eliminate past behavioral
habits. (Cameron used shock therapy, known as electroconvulsive
therapy or ECT, much more frequently than was accepted in that day,
more often than had been applied before in a clinical situation. Between 30 and 150 treatments—each treatment consisting of six shocks
in a row, twice a day—were given to each patient, a far greater fre-
quency of shocks than was the norm.) Certain of his patients—unknown to themselves or their families—were kept in a coma for
eighty-six days. Then Cameron and his associates reconditioned the
heavily sedated patients through psychic driving, exposing them to
tape-recorded messages played over and over from speakers under
their pillows. Some heard the same message a quarter of a million
times. The messages were simple but geared for maximum psychological effect. (They were the opposite of subtle: one patient recalled hearing "You killed your mother" over and over.) After many rounds of the
negative message, a positive "get well" message was substituted. By this
third stage in the treatment, most of the patients were no longer able
to perform even basic functions, requiring training as if they were ba-

bies in order to eat, use the toilet, or speak. Cameron explained, "The third stage of depatterning is reached when the patient loses all recollection of the fact that he formerly possessed a space-time image which served to explain the events of the day to him."[32] Needless to say, most memories were erased, and although some were retrieved as the patient went back through the stages in reverse order, it could not be predicted which ones.

By the end of the treatment, the patient had been duly repatterned. In Cameron's account, this meant that the patient's unconscious patterns of response, which were seen as the result of past traumas or childhood experiences, had been wiped clean, and new patterns had arisen to take their place. Whatever "schizophrenia" had existed before treatment was said to have disappeared, along with much of the rest of the personality. Despite objections, though muted, from some colleagues and from the overworked nursing staff at the facility, who frantically tried to care for their moribund charges, results appeared positive and were if nothing else dramatic: people emerged from the ward walking differently, talking differently, and acting differently. Wives were more docile, daughters were less inclined to histrionics, and sons were better behaved. Most had no memory of their treatment, much less their lives, and were, at least at first, grateful to their doctor for his help. A woman who had been admitted for mood swings and mild depression said about her release, "He was my doctor, and I thought he had cured me," even though she had no memory of her three children.[33] Despite the seeming cure, however, many ex-patients suffered severe functional and emotional problems with their families and work in succeeding years, and some, lacking a narrative of their previous lives, were left with persistent traumatic memories of Dr. Cameron's treatment itself.

"Behind closed doors," a Canadian/BBC-TV program introduced its lead story in 1998, "human guinea pigs in shocking mind control experiments conducted by government and the CIA." The Canadian and U.S. governments, along with the CIA, had not only provided financial support for Cameron's experiments but encouraged them and fostered the environment that made them possible. All of which leads to a further question: how did a progressive psychiatrist with a national and international reputation, who was recommended to patients far and wide as "the best," become fodder for a lurid-sounding telejournalistic exposé forty years thence? His patients and colleagues eventually rebelled, a group of the former uniting to sue the U.S. and Canadian governments for funding the research (the lawsuit was settled out of court), a group of the latter emerging after Cameron's death to criticize his methods.

The head of McGill University's psychology department during Cameron's reign over psychiatry there recently went on record to say, "That was an awful set of ideas Cameron was working with. It called for no intellectual respect. If you actually look at what he was doing and what he wrote, it would make you laugh. . . . Look, Cameron was no good as a researcher. . . . He was eminent because of politics."[34] The politics were twofold: the politics of constant maneuvering and finagling (characteristic of university departments and institutions), and the politics of superstates with their own imperative to preemptively create new weapons in controlling minds and behavior before the enemy did. American intelligence officials did not hesitate to "outsource" any basic research they felt was necessary. Thus in professional contexts, it appears, Cameron's work was tolerated even by those who would otherwise have objected. He lectured widely, published a lot, and received few personal or professional challenges while he was in his prime.

Like Wolff, Cameron was not an easy man to contradict. He was pushing hard, convinced his methods would bear fruit. By 1961, however, the research climate as well as the social climate was changing. Internally, several of the CIA's more outlandish experiments had gone dangerously wrong; moreover, the civil rights movement was bringing about a concomitant rise in consciousness of the rights of human subjects in medical and psychological experiments. Most of Cameron's funding was terminated. Three more years of experiments caused him to admit that his technique had proved useless: "a ten-year trip down the wrong road," he called it then, managing to destroy most of his patients' files. However, he did not offer an apology to his experimental subjects; nor did the Canadian government, the American government, or the CIA. A disappointed man, his reputation in decline, Cameron suffered a heart attack and died in 1967 while hiking in the mountains.

DOCTORS WEST, WOLFF, AND CAMERON were three of many scientists who used the principles of behavioral conditioning to bring about dramatic transformations in research subjects. Sociologists, anthropologists, psychologists, psychiatrists, and neurologists went "further" under the aegis of Human Ecology, MK-ULTRA, and satellite projects than they otherwise might have done. Well-intentioned anti-Communist scientists of a decent sort produced indecent outcomes. The government, in its efforts to fight terror, got very good at sowing it.

In elite professorial circles during the 1970s, it suddenly became necessary to defend one's work for the CIA. Such work became an incendiary topic, touching off hot debate among those whose careers had been active during the 1950s and 1960s. The necessity of justification quickly became the justification of necessity, and many researchers stressed that back then such work had been more in the heroic vein than any other and certainly nothing to be ashamed of. On the other hand, several Human Ecology–funded researchers stated that, had they known they were working for the CIA at the time, they would never have accepted the money. (Sociologist Jay Shulman and psychiatrist Carl Rogers were among this group.) A few, discovering whom they'd been working for, were mainly quizzical: "Why were they backing me? What were they getting out of this? I still don't know," said an anthropologist who studied cultural accommodation.[35] And finally, once MK-ULTRA and other programs became the subject of a 1977 congressional hearing, certain CIA scientists wanted to make sure the professors didn't come out looking like the head-in-the-clouds victims of a nefarious agency: "Don't get the idea that all these behavioral scientists were nice and pure, that they didn't want to change anything, and that they were detached in their science," a CIA Technical Services scientist warned. "They were up to their necks in changing people. It just happened that the things they were interested in were not always the same as what we were." Human engineers had always dreamed of enacting enormous alterations of the human body and mind through environmental and other manipulations. As it happened, this long-standing aim received a boost from others who had their own reasons for dreaming of enormous changes. Spies and scholars, with their own distinct visions of an imminent utopia, proved capable of sharing research methods and results.

Consider the case of the remote-controlled cat. Trained by MK-ULTRA–funded scientists in pure behavioral conditioning, it could sidle up to enemy agents on city streets or at ill-lit assignation points and record their conversations by means of a surgically implanted microphone. A lot of work went into the animal—multiple surgeries to implant remote sensors and listening devices and to control its peregrinations. A last addition was a bomb that could be detonated at the push of a button. After much training and expense, the cat went out on its first exploratory mission but was run over by a car while crossing the street. Likewise, dolphins trained for illicit purposes did not work out as planned. By this point an almost hallucinatory quality had entered into these experimental sallies. Yet these and other MK-ULTRA programs continued for another couple of years, at least into the mid-1960s.

In a sense, the experiments showed something hopeful: that one could not simply brainwash a person and be done with it. The grandest Grand Guignol goals failed, for even in the case of the utterly reduced and reprogrammed, the wiped-clean subjects, the surgically altered, human and nonhuman experimental subjects alike, a perfect control could not be achieved. "All experiments beyond a certain point always failed," said an MK-ULTRA veteran, "because the subject jerked himself back for some reason or the subject got amnesiac or catatonic."[36] Or was run over, or ran away. Certainly one could create a "vegetable," but this was not useful. Having arrived at this disappointing yet in some ways cheering knowledge, MK-ULTRA itself was discontinued in 1963, although research continued for some years through the Science and Technology Directorate. More successful and more significant was the CIA scientists' contribution to a method of subtle control, perhaps better described as a persuasive and all-pervasive conditioning that proved well-nigh inescapable for the very reason that it was experienced as rather nice.

In the geopolitical sphere, these mind-control and behavior-control initiatives had long-lasting effects. They formed the basis of CIA and Special Forces methods for training foreign armies in an array of cold-war theaters. During an energetic and munificently funded period from the 1960s through the 1980s, scientific advances in behavior modification were exported to America's Latin American backyard, to Vietnam's battlefields, to the Asian "tiger" states, and elsewhere. West Point required its students to take courses in human relations, military psychology, and "special warfare" (psychological warfare and forceful interrogation techniques). The CIA-backed School of the Americas taught Argentine, El Salvadorean, and Panamanian soldiers counterinsurgency techniques for rooting out subversives, imprisoning people, and forcefully extracting information, with the result that "proxy armies" were "standardized."[37] A behavioral science approach to the "cultural engineering" of emerging Third World states was increasingly accepted. A Vietnam-era program, the Special Operations Research Organization (known as SORO), for example, provided the army with dozens of country-specific handbooks on psychological operations, case studies to aid in exploiting the psychological vulnerabilities of Southeast Asian people, and a comprehensive databank of behavioral information called the Counter-Insurgency Information Analysis Center.[38]

The specter of Communist brainwashing capabilities haunted national intelligence and strategic agendas for years during the cold war, and this fear supported the work of scientists who claimed to have access to the territory where brain, body, and mind met. The speculative and spectral began to seem real. Dr. West, for one, linked his vision of a truly unified behavioral science that combined insights from neuroscience, pharmacology, electronics, sociology, psychology, and cultural anthropology to a dream of access to total knowledge: he spoke of building a massive databank of consciousness itself, which would consist of bioelectric recordings—gathered through electromyography, electro-oculography, rheoencephalography, and CAT scans—subsequently analyzed and coded, then stored in a centralized location such as the National Institute of Mental Health or the National Medical Library. The dreams of knowledge and power that had fueled human engineering for decades were propelled forward.

IN A LAST, SURPRISING TWIST of the CIA's behavioral science program, Wolff, West, Cameron, and others set out to make a true science of human behavior. Their arduous research investigations testify, more than anything, to the human vagaries and behavioral mysteries inherent in this pursuit. Once again the strangest activity displayed was not so much that of the cats, rats, dogs, elephants, refugees, drug addicts, hippies, schizophrenics, housewives, soldiers, and other subjects the scientists studied, but that of themselves. The curious fact is that they were the most exorbitant subjects of all in these most exotic of experiments.

CHAPTER 11

The Real World

TWO PROJECTS, both from the early 1960s, marked a turning point for the experimental impulse toward human engineering that had grown throughout the twentieth century. At Yale the infamous Milgram experiments on obedience tested how likely ordinary people would be to inflict agonizing pain on another when ordered to do so. (On average, they were quite willing.) At Harvard the notorious Leary experiments with psilocybin explored how inclined prisoners in a maximum-security jail would be, after taking a mind-altering drug, to change their habitual ways of thinking and acting. (On average, they did change, although not permanently.) In short, the experiments traced opposite speculative arcs: one turned normal, decent people into torturers, and the other turned hardened criminals into generous human beings concerned for the welfare of others.

Considered jointly, these two sets of experiments took place at the very limit of what was ethically and legally possible within an American laboratory. Neither would be permissible today under current guidelines for research using human subjects. To study these projects now is to turn a microscope on how the impulse to experiment with human materials under controlled conditions reached a critical point and, as a result, was rather abruptly altered. Even as one set of possibilities died, they were reborn in the America of the 1960s. By tracing the conditions, designs, and results of these two controversial undertakings, we can observe yet another process by which experiments in human engineering left the laboratory and entered the world at large.

Loosely, the two sets of experiments moved in two different directions, which have been perpetuated in the twenty-first century. One moved toward paranoia: the splitting, fracturing, and torturing of consciousness, the disorganization of mental-physical patterning. The other moved toward metanoia: the expanding, unraveling, deconstructing, and reconstructing of consciousness, the reorganization of mental-physical patterning. Most people know what *paranoia* is; in common parlance it means excessive suspicion of others or delusions of grandeur in oneself. One feels persecuted, a feeling that comes from an inaccurate assessment of the situation in relation to the self. The Greeks used it as a word for madness. *Metanoia* is a term less known and has no real dictionary definition, but it can be defined provisionally as what the Irish writer and visionary Gerald Heard called the "growing edge," a collective change in thinking or consciousness. Paranoia and metanoia: each is susceptible of being multiplied within different experimental environments.

SOCIAL PSYCHOLOGIST Stanley J. Milgram went to an extreme in testing the willingness of average people to obey orders. In 1961 he set up a laboratory study at Yale that has become perhaps the most famous psychological experiment of the twentieth century. Milgram placed a newspaper ad announcing his search for subjects to participate in an experiment on memory. Volunteers were paid four dollars (not much, even then; most of the people who responded said they were being civic-minded). They entered an imposing building on the Yale campus and reported to a second-floor room, where they were paired with a "dummy" partner. After brief introductions, the partners drew straws to see who would be the "learner" and who the "teacher" in the experiment. The outcome was rigged, and the dummy partner was whisked off to an adjoining room where he could not be seen but, importantly, could be heard. Next, the subject—now designated the teacher—was told to take a seat in front of a large machine equipped with a lever. It was designed, he was told, to deliver a range of electrical shocks, the intensity of which was marked on a front panel. Reading out a simple word test, the subject was instructed to deliver a shock via electrodes attached to his partner's arm whenever the answer he gave was incorrect. A brief visit to the back room allowed him to watch as a technician smeared electrode paste on his partner's arm to ensure the "full contact" of surfaces. For purposes of gauging the shocks, the subject was then given a shock from the lower end of the scale on his own forearm. Some people gasped, for the shock was stronger than they anticipated. After these preliminaries, the test proceeded.

Although the participants believed the researchers were testing their partners' performance on the word test, it was in fact their own performance that was being tested. When the partner gave a wrong answer—for he had secret orders to do so—the subject was told to increase the voltage of shock he administered. A white-lab-coated Yale scientist with a clipboard stood behind him in the control room, urging him on with a nod or, when he hesitated, a few terse words along the lines of "The experiment requires that you continue" and "You have no choice but to continue." Screams and pleas for mercy began coming from the adjoining room. There was pounding on the wall, and the partner in the other room could be heard yelling, "Get me out of here!" and complaining of a heart condition. (In fact, his cries were pre-recorded and played on cue from a tape recorder sitting next to him.) Yet the overwhelming preponderance of participants—on the order of two-thirds—continued to administer shocks that they believed to be approaching a lethal level.[1] These results were soon to dismay audiences both professional and popular with an icily dim view of the ability of one man to feel the pain of another. In point of fact, dismay is not the only emotion the results aroused in onlookers; something akin to glee can also be discerned, but this particular we will leave to address later.

Here it becomes necessary to introduce a modicum of information about the scientist who "masterminded" these experiments. Stanley Milgram was twenty-seven and had just gained a post on the tenure track of the Yale psychology department where, although an up-and-comer, he was still on tender footing. For Milgram was not quite the sober social scientist through and through: many accounts mention that he was at heart an artist who wrote librettos in his off-hours, loved the stage and dramatic conflict on a biblical scale, and dreamed that his creative writing would someday find a readership. In addition, he was a Jew from the Bronx, a practical joker, and occasionally lacked gravitas: not an auspicious background at a university known for its persistent Skull and Bones lineages. But his "obedience" experiment made Milgram famous the world over. Much debated and much mentioned, his lab work occasioned a play, a TV drama, a movie, and a great deal of reflection among the morally inclined (Thomas Merton discusses the experiments in *Raids on the Unspeakable*), and it took a central place in Holocaust debates and debates over what used to be called "human nature" in general. The one thing it did not do was give Stanley Milgram tenure.[2]

At the time, Milgram was perfectly willing to tap the dramatic potential of his work. Film footage of the experiments is still used in introductory psychology classes at Yale and elsewhere: in grainy black and white one can see the participant sweating, hesitating, turning with an imploring look to the official-looking scientist standing behind him, then returning to the large apparatus and pulling a lever at the end of the range. The camera pans slowly across the machine's scale, from mild to strong to extremely strong, and finally to a point marked "Danger: Severe Shock" (which always gets a laugh out of undergraduates today, perhaps because the technology looks outdated or because the film's technique of the "slow pan" is so hokey). Some participants refused to continue beyond a certain point, and these performances are also captured on film. The first one was a man in a lumberjack shirt and heavy-framed glasses, not highly educated, an average New Havenite blue-collar worker. "The experiment requires that you continue," the scientist tells him. "I don't care what the experiment requires, I'm not going any further," says the man, and lights a cigarette. Most people, on the other hand, continue to follow orders. Some go to the very end of the scale and, told to start again from the beginning of the word test, keep zapping at the high end until instructed to stop.[3]

Then and now, the experiment seemed to dramatize anew Hannah Arendt's observations about the banality of evil. If ordinary New Haven residents were willing to go to such extraordinary lengths, indeed to torture their fellows, over a manifestly trivial word test, then it surely was not hard to explain why extreme stress situations brought about murderous behavior in a well-behaved population. In fact, Milgram deliberately based his experiment on *Eichmann in Jerusalem*, Arendt's account of the 1961 trial of the notorious Nazi. Attending Eichmann's trial, Arendt had been struck by the character of this man who "with great zeal and the most meticulous care" arranged the transportation by train for some five million Jewish men, women, and children out of ghettos and homes in Germany, Eastern Europe, and Western Europe to Buchenwald, Auschwitz, and Theresienstadt. Expecting him to appear a monster, Arendt was surprised to discover on the stand a nondescript man who did not even distinguish himself by a special dislike for Jews. Eichmann had had a Jewish mistress, and he tested well on psychological tests: a family man, perfectly normal ("more normal, at any rate, than I am after having examined him," remarked one psychological expert).[4]

The most striking thing about Eichmann, for a supposed criminal mastermind, was his dullness. He spoke only "bureaucratese" (the German word *Amtssprache* was one he used himself) and was incapable of uttering a sentence that was not a cliché. Arendt argued that his formulaic approach to language allowed him to protect himself against the existence of other people—in short, it allowed him to compartmentalize his work. Having no stomach for killing or for gruesome details, he made sure to carry out his tasks without considering any hint of their consequences, even when they were thrust upon him. (At one point he was forced to visit a death camp and fainted at the sight.) In the early days of the Final Solution he actually tried to save the lives of a shipment of Jews by diverting their path of deportation, a step that caused him trouble with his colleagues and that he never again attempted or even particularly recalled, leading Arendt to conclude, "Yes, he had a conscience, and his conscience functioned in the expected way for about four weeks, whereupon it began to function the other way around." Nor, on the other hand, was he above bragging in bars of having brought about the "death of five million," once he was safely out of Europe and resettled in a small town in Argentina. To Arendt, Eichmann was monstrous precisely because he was not a monster. A "normal" human being had done the unspeakable—and spoke of it in lamentable clichés. One interpretation was that he suffered from an excess of mechanically triggered blind obedience.

Moved and fascinated by Arendt's controversial thesis, Milgram decided to test whether mechanical obedience could be elicited under more putatively democratic conditions. Thus the Milgram experiments are sometimes called the Eichmann experiments (although Arendt vehemently denied any connection between the two).

The Milgram results were notorious enough, but a backlash fed their notoriety further. The participants, it seemed, had not blithely obeyed orders but had suffered nervous breakdowns and other forms of psychological and physical stress following the experiment. "Many subjects showed signs of nervousness in the experimental situation, and especially upon administering the more powerful shocks. In a large number of cases the degree of tension reached extremes that are rarely seen in socio-psychological laboratory studies," Milgram reported.[5] They hadn't *liked* doing it, a fact that Milgram had interpreted as further evidence of authoritarian inclinations. They were not sadists, but that was somehow worse, argued Milgram. Why hadn't they simply "broken off . . . as their conscience dictated"? Was there a hidden Nazi in two-thirds of the good citizens of New Haven? Some deemed the experiments unethical, but his results endure as a powerful form of folklore, useful in explaining situations in which a person suddenly finds herself doing something untoward.

In 1974 during the Patty Hearst trial, Milgram's sheeplike subjects were mentioned as a prominent laboratory example of the "attitude change" to which the Symbionese Liberation Army subjected its kidnapped heiress. A recent article in *The New Yorker*, "Why Do People Follow Dictators?" mentions Milgram's experiments as evidence of an authoritarian inclination in all of us, confirming Max Lerner's remark that "in everyone there is a little bit of the fascist and in some people quite a bit."[6] Finally, Milgram received mention once again to help explain how ordinary GIs from Cumberland, Maryland, could have been so brutal—and so seemingly casual—in their treatment of Iraqi prisoners at Abu Ghraib. Milgram's remain some of the most famous or infamous psychological experiments ever conducted in America.

IN CONTRAST, the Harvard experiments at Concord Prison in the early 1960s, under the auspices of the Prison Project, were meant to be cooperative instead of coercive, and to tear down any vestige of a barrier between scientist and subject. Timothy Leary came to Harvard in 1959 at the behest of David McClelland, a leading member of the by-then well-established Department of Social Relations, a prominent proponent of the unifying approach to the social sciences typical of the laboratory imagination. He hired Leary for his promising "existential transactional" approach to the field. "Existential" meant that the scientist would get out of his laboratory and into the world, and "transactional" meant that he would meet his subject on an equal footing. (Leary had developed this approach in California, where he worked for the Kaiser managed-care conglomerate as one of the first psychologists to design diagnostic tests—tests later used on him when he was jailed in San Quentin for possession of marijuana.) In 1960, at the age of thirty-nine, he took his first dose of magic mushrooms while vacationing in Cuernavaca, with the McClellands just down the road as neighbors. Leary had visions of the mind as a thirteen-billion-cell computer, a reducing machine, a huge filing cabinet, and a "repetitious narcotic," which he related with some excitement the next day to McClelland, who was distinctly skeptical.

Back at Harvard, while artists and philosophers such as Robert Lowell, Arthur Koestler, William S. Burroughs, and Allen Ginsberg were taking part in psilocybin sessions (Thelonious Monk had it delivered to his door), Leary devised the Prison Project to run at nearby Concord Prison. The novel idea was that social scientists would *sit with* the prisoners, undergo the experiment with them, and take the same drug. Therein lay the ethical core of the Prison Project: experimenting with, not experimenting on. (The story of how Leary got Concord Prison officials to agree to let him "turn on" with prisoners for two years is unusual but not to be dwelled on here. Suffice it to say that it was a tribute to the power of the Harvard imprimatur to open any door and the ability of Leary to maneuver through it.)

So it happened that at the same time Milgram was conducting his obedience experiments at Yale, Leary began his psilocybin experiments through Harvard. The first test took place in a ward of the prison infirmary on the morning of March 27, 1961: Leary and three prisoners took psilocybin pills while two graduate students and two other prisoners acted as guides. To counter the prison's dreariness, Leary brought along a record player and several picture books (including *The Family of Man*, Edward Steichen's famous collection of photographs of the universal human story), but as the pills took hold, nobody felt like listening to music or looking at pictures. As Leary recalled, "I felt terrible. What a place to be on a gray morning! In a dingy room in a grim penitentiary, out of my mind." The man sitting next to him, a "Polish embezzler from Worcester, Massachusetts," looked equally grim, his pores frighteningly large, his skin bad, his teeth decaying in "green-yellow enamel." What happened next is worth quoting from Leary's own account:

I said to him with a weak grin, How are you doing, John? He said, I feel fine. Then he paused for a minute and asked, How are you doing, Doc? I was about to say in a reassuring psychological tone that I felt fine, but I couldn't so I said, I feel lousy. John drew back his purple-pink lips, showed his green-yellow teeth in a sickly grin, and said, What's the matter, Doc? Why you feel lousy? I looked with two microscopic retina lenses into his eyes. I could see every line, yellow spider webs, red network of veins gleaming out at me. I said, John, I'm afraid of you. His eyes got bigger, then he began to laugh. I could look inside his mouth, swollen red tissues, gums, tongue, throat. I was prepared to be swallowed. Then I heard him say, Well that's funny, Doc, 'cause I'm afraid of you. We were both smiling at this point, leaning forward. Doc, he said, why are you afraid of me? I said, I'm afraid of you, John, because you're a criminal. He nodded. I said, John, why are you afraid of me? He said, I'm afraid of you, Doc, because you're a mad scientist. Then our retinas locked and I slid down into the tunnel of his eyes, and I could feel him walking around in my skull and we both began to laugh. And there it was, that dark moment of fear and distrust, which could have changed in a second to become hatred and terror. But we made the love connection. The flicker in the dark. Suddenly, the sun came out in the room and I felt great and I knew he did too.[7]

Word spread among prisoners about the good drugs the Harvard professors were giving out, and inmates vied to try them. One of the participant-scientists, then a graduate student at Harvard, described what happened in the Concord Prison as follows: "The prison walls were down; the whole world was wide open."[8] Far from suffering nervous breakdowns and being decimated by stress, as Milgram's subjects were, Leary's prisoners seemed uplifted. His results overwhelmingly showed that even inveterate criminals changed their behavior as a result of the experiment. Few returned to crime after release from prison, he claimed. (Leary fudged some of his results about recidivism rates: although psilocybin was revelatory, its effects lasted only as long as the bonds between the participants did. Once back on the streets, many returned to doing whatever it was that had got them in prison in the first place.) Still, even if only for this one moment, the usual arrangement between scientist and subject or between expert and object of knowledge was radically altered, or so they felt. The Prison Project ended when Leary, along with his colleague Richard Alpert, was dismissed from Harvard. (As both were proud to say, the only Harvard professor to suffer summary dismissal before them was Ralph Waldo ern faith in direct ability to change people.) At some point, however, through the use of these drugs, he was deemed to have lost the capacity to be a proper Harvard social scientist. Aldous Huxley, who happened to be spending a semester at MIT in 1961 when the prison experiment began, had this to say about social scientists, in particular those who try to explain human consciousness: "They talk about what they know. But what they know isn't worth talking about."[10]

Once Leary believed he had had direct experience of full reality or expanded consciousness or whatever one calls it, the scientific paradigm was undermined: he spoke of something not contained in a laboratory, at least not in the kind of laboratory social scientists in his tradition had built. If scientists heretofore had used imagination to strengthen their ability to exert a certain kind of control, then the new breed of radical experimentalists—wall-breakers, paradigm-shifters, consciousness-explorers—wanted to use the laboratory to permit the breaking of control, the shifting of perspective, the altering of models, to undermine ingrained habits and the "prolonged mental rutting" that Marshall McLuhan for one had said was characteristic of life in the mid-but-becoming-late twentieth century. Leary as well as some others of his cohort who remained within the universities saw a truth about the laboratory that had not been evident to earlier experimentalists: it replicates its own conditions in the form of "results." Scientists not only tend to find confirmation of their preexisting worldviews—after all, they are working in an experimental "world" of their own design, where they are in charge of setting experimental conditions as they will—but they also alter the world in the process. Running such experiments causes marked and rapid changes to occur both inside and outside the laboratory. Meanwhile experimentalists may imagine they are "standing apart" and observing their subjects impartially, but they are always a part of the reality they are altering.

Indeed, Harvard's department and Laboratory of Social Relations—Leary's academic home until his 1963 dismissal—was a bastion of this style of observation. In the decade before Leary arrived there, a key

Emerson in 1834.) Both went on to become counterculture fig
heads. Alpert took the name Ram Dass and went off to meditate in
dia, while Leary combined freewheeling drug use, political activi
and visionary preaching to achieve the status, in Richard Nixon's v
at least, of "the most dangerous man in America."

The aim of the Prison Project was to dismantle the wall between s
entist and subject. If the Milgram experiments flooded the laborato
with a powerful if not entirely understood authority, the Leary exper
ments tried to sweep authority from the laboratory entirely ("I coul
feel him walking around in my skull"). If Milgram's experiments mad
the world into a prison, Leary's made the prison into a world. But in
dismantling this wall, the professor—more than even he realized at the
time—shifted from scientific to mystic ends, privately saying, "Let's see
if the convicts can become Buddha." In this respect, it is important to
note, he did not stray from the original impetus of the laboratory imagination as traced in this book: from the beginning, his aim was to use
science to settle the age-old questions that philosophy had always addressed, questions about humans' fullest capacities in relation to others,
the meaning of consciousness, and the contours of the reality to which
that consciousness had access. As a purple-prosed *New York Times* editorial described the Yale Institute of Human Relations in 1929:

> Here is to be brought into a compendium of wisdom all that the varied
> natural and social sciences have learned about this adventurous piece with
> which Nature has crowned her creation—a cosmography of the human be-
> ing within whose body, which holds the mind as its guest, the greatest won-
> ders of the world are held but little known in their relations, despite all that
> has been said and written of them.[9]

The human being as a microcosm, with access to the wonders of the
world—this Renaissance vision of hope-for-understanding joined to a
twentieth-century hope-for-control was one Leary shared. (When
Leary spoke of the insights psilocybin or LSD granted him and the
Concord prisoners, he combined faith in knowledge with a more mod-

project in the laboratory took human engineering to new extremes: a "special room" for the study of small groups, the creation of a sociologist named Robert Freed Bales. Although Bales was laboring, academically speaking, in a subfield of a subdiscipline of sociology proper, his work aspired to great generality: he hoped to construct a laboratory version of how human social bonds are formed and function, and to be able to quantify this process in terms of numerical "phases," measured, charted, and graphed. Thus his work was not reducible to any single social-scientific field but held the promise of bringing them all together: it was, Talcott Parsons said, "peculiarly favorable" for the development of the ultimate grand theory and, for a while, was at the forefront of hopes for human engineering.[11] Through this work social bonds between people could at last be studied in a properly scientific manner in a properly controlled laboratory.

Bales proceeded by constructing a space in which a social scientist could observe more closely than ever before the way human beings interrelate. In an unexpected development, however, during the late 1950s Bales's work began to fade in importance, and it made its most lasting mark as the basis of the modern-day focus group. (Anywhere consumers' or citizens' attitudes are being measured in "focus groups," one can find in use Bales's architecture of one-way mirrors, adjoining observation rooms, and recording devices, although the equipment has been updated.) In 1963, when Timothy Leary was ousted from Harvard, one of his parting exchanges was with Bales in the laboratory they had shared: turning to leave, Leary according to legend asked, "Freed, do you think I'm psychotic?" Bales pondered the question for a moment and replied, "Tim, I just don't know."[12] Momentum was with Leary, however, for a new kind of human engineering laboratory was emerging outside the confines and redoubts of the old one.

Leary's work at Harvard marks a critical point in the social or human engineering trajectory. Its exorbitant strangeness became too much for those who preferred an earlier style of laboratory "control." Even so, the quasi-mystic role of philosopher-king that he took on was present if latent in all the work of the laboratory imagination, a strange near-mystical impulse welling up within the putatively closed and determinedly rational space of a machine-based and experimentally defined laboratory.

WHAT DID THESE TWO SETS of experiments prove? On the one hand, they proved that a person can be reduced to a morally negligible near-automaton rather quickly in an afternoon's work—even in relation to the person's *own* standards. Comfortable assumptions such as "I am not the kind of person who would torture another human being when ordered to do so" were no longer tenable. On the other hand, they showed that a person's ability to perceive the world in a certain way can be refigured. From total control to almost no control was a somewhat dizzying reversal.

Ever since Darwin, ever since Locke, and probably going back to the Pelagians (believers in free will over determinism) thinkers have debated the extent to which human beings are controlled by conditions and circumstances, and the extent to which they choose to act free of those conditions. But importantly, these twentieth-century experiments were conducted in the controlled space of the laboratory. What value, then, did they really have?

Milgram created a laboratory version of an eternally recurring scenario, for as he said, "The situation in which one agent commands another to hurt a third turns up time and again as a significant theme in human relations."[13] From God commanding Abraham to kill Isaac to more current examples of a squadron leader telling a seventeen-year-old to kill Vietnamese civilians, a panhuman phenomenon exists in which X tells Y to hurt Z, and Y carries it out. Milgram felt he had isolated this phenomenon in the laboratory; he had captured it, so to speak. He certainly was right about obedience to authority—but the most interesting part of his experiment is an unexamined and difficult-to-see aspect of the experimenter's own authority.

Milgram's experiment separated the participant from his supposed partner as much as it separated him from his superior, the Yale social scientist. The subject was completely isolated in a made-up world where, unknown to him, everyone else was acting a part. Imagining himself a mere cipher for carrying out an experiment on the mnemonic capacities of his supposed partner, he was in fact the cynosure of the experimental eye, the core of what was being tested. The entire situation was rigged to ensure the utter isolation of the subject. It was as if an unwitting looker-on stumbled onto a stage to find himself playing a starring role in a drama in which he had never agreed to act and did not even know was under way. When the house lights came on at last and the subject saw the key part he had played, he blinked and found himself in the throes of a humiliation too great even to articulate to himself. We the audience cannot help but say, "Serves him right, the Nazi."

And of course, this was precisely Milgram's point, to test obedience to a compelling authority. But even though the experiment was about an authoritarian situation, the giving and following of orders, it dramatically showed something else. Built into the design, and the *conducting of it* in the name of anti-authoritarianism, was a playground joke, like an episode from *Candid Camera*, which in fact was Milgram's favorite TV show. The real drama occurs not when the subject administers the electric shocks but in the last act when the experimenter informs the subject what was really going on. This phase was dubbed in an early publication "dehoaxing" and in later publications renamed as the more professional-sounding "debriefing." Almost always the beneficiaries of debriefing express great relief not to have been hurting the person (whose cries for mercy they had only minutes before heard issuing from the adjoining room). Immediate and often pathetic attempts to justify themselves ensue. Subjects writhe. Some start talking about how worried they were about the guy, others about how much charity work they've done or how much they love their families. They are exposed at the moment of understanding that it was they who were being tested and observed all along—and usually not to great advantage. (One Milgram participant recalled, over forty years later, "They kept saying, 'You didn't hurt anyone, don't worry, you didn't hurt anyone,' but it's too late for that. You can never really debrief a subject after an experiment like that. You've given shocks. You thought you were really giving shocks, and nothing can take away from you the knowledge of how you acted. There's no turning back.")[14]

The peculiar form of power created when one reveals one's trick afterward comprises a significant part of the urge to experiment. Here motivation is key: the experimenter must be invoking very high social goals in order to justify behavior that otherwise might seem duplicitous or sadistic. The highest social goals in this case, as in many, are Experimentally Derived Knowledge and Science, the Greater Good of Understanding Human Nature, and Teaching People a Lesson (What They Are Really Like).

What Milgram and Leary and their illustrious if forgotten predecessors prove is that people can indeed be reduced to things (unthinking mechanical operatives, able to carry out military orders or social imperatives or dull jobs)—not always and not everyone, but a significant proportion of the time and to a significant degree any person can, and this occurred in the Milgram experiment. It happened not to the man apparently being shocked (the supposed "object" who is actually an actor sitting comfortably behind the scenes playing taped messages of himself screaming) but to the Average Joe who believes he is torturing someone. Another way of saying this is that it is the torturer, not the tortured, who is dehumanized. It also happened to Milgram himself. For to turn this transformative capability into a parlor trick, as Milgram did, is to toy with people.

In the event, the only people easily held up for disdain are the New Haven citizen-volunteers; Milgram's own role is more difficult to see. His experiment, which like all experiments is a snapshot of the circumstances in which it was performed, displays the workings of a historical logic that continues to operate: a logic that blames "bad apples" for crimes that each person commits or condones in small and large ways every day. Furthermore, Milgram's experiments helped perpetuate these conditions. He created a paranoid laboratory space that made paranoid actors who in turn acted to bring about a paranoid world—broken, that is, from reality as it was and yet conforming, somehow, to reality as it then became and would continue to become until the cycle was changed.

For some reason, this phenomenon—call it the coercive dynamic—
was and remains hidden even when it is in plain view. Very few people
have been able to see that Milgram was partaking of exactly the attitude
for which he was castigating others. "An aspect of the research *setting*
had become part of the research," the psychologist Benjamin Harris has
concluded, and therefore the experiment was not simply ethically com-
promised but methodologically confounded.[15] The guilty participants
could not answer back, having been revealed as moral ciphers; and the
casual observer, accepting the results at face value, rarely did. Just so, in
a series of experiments conducted at Stanford University in 1971,
groups of undergraduate volunteers were kept in a makeshift "prison"
and systematically humiliated (hooded for long periods, kept naked,
kept sleepless, interrogated to the point of breakdown, etc.) by their
peers. Many onlookers—visitors to the lab, including the participants'
parents, a legal expert, and a priest—merely accepted the manifestly
unfair psychological-torture-in-the-name-of-science that was going on
in the basement of the psychology department. (Very few people have
perceived these experiments as anything other than stark revelations of
certain truths about human nature, truths that, however, are rarely applied to oneself. Once again, much derision was directed at the hapless participants, along with assurances that such an experiment as
Milgram's or the Stanford prison, were it even permissible to perform
today, would never get the same results now as people are less "conformist" than they once were.) The totalitarian streak so easily blamed
on the compliant subject was native to and generated by the author of
the experiment and the very structure of the experimental conditions
he imposed.

The Milgram experiment, which is often described as "classic," was
about an inescapable and unprotestable coercion. Milgram used an extreme form of control to argue, in essence, against controlling structures. He degraded and humiliated his subjects in order to edify and
advance knowledge of the human capacity to degrade and humiliate
others. He used the ignorance of others in an experimental situation to
prove to them that they are ignorant of their ignorance. The experi-

ment speaks volumes about the impulse for scientific experimentation in the laboratory and control in the world.

WITH MILGRAM AND LEARY, human experiments were in a real sense operationalized. Every department store, every classroom, every enclosed social situation, every personal encounter has become a potential laboratory. (Or an actual one, as witnessed by the presence of social science techniques in your neighborhood Gap and Starbucks, among other places, where employees or statisticians carry out observations, record data, and quantify results, all in an effort to modify patterns, change behaviors, monitor attitudes, and deliver gratification.)

The Milgram and Leary experiments coincided with a growing trend to extend experimental arenas by imposing controls on the surrounding environment. The laboratory, in a sense, became wider. Even as the use of human subjects was contested, questioned, regulated, and refined (in part as a result of the Harvard Medical School researcher Harry Beecher's 1966 exposé showing exactly how run-of-the-mill, and therefore alarming, the abuses then condoned in experiments using medical patients as subjects were), the experiments themselves transmuted. As the inner space of the laboratory became more and more subject to outside scrutiny—Watson's and the CIA's experiments, for example, prompted horror and disbelief—researchers began to conduct experiments in an array of other spaces. Quasi-experimental spaces such as focus groups, survey interviews, fieldwork sites, "retail anthropology" devoted to the intensive recording of patterns of consumer behavior, and even the widespread use of Handycams in public and private life have brought the experimental eye into a variety of social situations. Just as Milgram's and Leary's procedures triggered concern for laboratory subjects, so too did they launch a new approach to experimenting, the experiment-as-theater, the take-home experiment, the *Candid Camera* eye with which we might always watch ourselves and others.

Most of the human engineers of the first two-thirds of the twentieth century wanted to observe human and animal subjects in highly artificial situations in order to change them in an "ought" direction. For example, the Yale group looked at rat peregrinations in order to dictate the "maze we all must learn"; the Hawthorne experiments were meant to optimize worker productivity and minimize worker dissent. Milgram, too, aimed to change people in a particular direction: he believed his work *forced* people to be more aware of social situations and thus more likely to question authority. (He was teaching them a lesson in order to change them, and he liked to quote a grateful letter from a participant who said being a Milgram subject caused him to question fighting in the Vietnam War.) By the early 1960s, when the so-called "experimental moment" came to the social sciences, the arts, and society more generally, these assumptions were overthrown. The goal became to see how things actually were in real life without immediately jumping to how things should be. The situation did not need to be constructed, and neither did the results. It was "is," not "ought." Also, the mode of operations changed: convicts and scientists shared the same experimental conditions in Leary's experiment, and he attempted to break down the separation between the scientist and his or her subject. By means of these changes in premises and modus operandi, some scientists felt, real change might occur. This is one reason why "the Sixties" is often seen as an experimental time.

For a time the experimental mode seemed to threaten to overturn the apple cart of middle-class norms and values. Experimental social scientists such as Leary, Alpert, and R. D. Laing questioned assumptions, undermined social givens, and suggested that merely living one's life according to the norms and demands one was born into was a form of brainwashing. And then this drive to experiment—adapted from the laboratory, loosed in the public sphere, handed out free on the streets—was transformed yet again. It became a refurbished ultra-deluxe greengrocer, the core of a new middle-class worldview premised on the consumer of life experiences. Whereas it once had seemed to offer a chance at changing how human society was run and how people perceived each other and their environment, it now offered a new status quo. For as it turned out, the experimental technique of observation that was practiced in the lab, tested under controlled conditions, operationally guided, and set free in the world had no particular aim built into it. It was neutral. As Susan Sontag pointed out, we see ourselves and others by means of "countless Webcasts, in which people record their day, each in his or her own reality show."[16] Out of the laboratory emerged . . . technique. But here is the crux of the matter: it was technique available to anyone, so it could always be used to make a point, to promote a greater idea, to carry on a crusade, to argue a cause, to fight a war.

THESE EXPERIMENTS show how activities confined to the closed space of a laboratory or field site are connected to the outside world in intimate ways. On the one hand, the experiments, like many that came before them, were able to "leave" the laboratory through several routes. The most obvious is the media, which employs as its very substance the techniques of human engineering first pioneered in laboratories, such as attitude influencing, mass polling, and the fine-tuning of information to modify social and personal goals. Advertising in and of itself is an ongoing experiment concerning the extent to which any creature can be conditioned—is there any limit to how many messages a person can be bombarded with in a day?—and at the turn of the twenty-first century, conditioning penetrates more and more surely into hitherto unreachable areas. Then too, as we have seen, the CIA pushed hard to operationalize the techniques of human engineering as applied to specific situations of interrogation and personality alteration, and it also spurred research into longer-term goals of mass attitude adjustment. The movement from laboratory to society happened all the time, but during the early to mid-1960s it reached a critical point and happened very fast. Or perhaps it happened in a different way: researchers, students, and average people protested the smooth importation of techniques for fine-tuning the self and soul and attempted to hijack these techniques.

This process of going out of the laboratory and into the world was also in some sense an illusion. Actions taken within a laboratory setting are also part of the world. They do not merely simulate reality, they *stimulate* it. One could argue that the false belief that experiments in the lab are somehow protected from real life—"it's just an experiment, it's not really real"—is what has allowed some of the more alarming recent abridgments of the lives of human beings and other creatures to occur. This belief, along with the assurance that such activities are going to bring about improvements in real life or protect the national security, has been enough to sanction the treatment of living subjects as inanimate objects. In some sense these experiments, any experiments, are *already* operational merely by being performed in a certain context, according to certain rules, and with certain goals as their intended outcome. What happens in small rooms has a large effect on the world.

CONCLUSION

THERE IS NOTHING quite so out of date as an earlier era's vision of the future. Yesterday's up-to-the-minute sci-fi imagining—from the *Italian Futurist Cookbook* to the *Jetsons'* space-age high-rises and gadgetry—is today's lovable kitsch. Likewise, there is nothing quite so odd as an earnestly logical, no-holds-barred attempt to Know the Universe, to divide it up in parcels, to devise forty-two equations for its workings, and ultimately to file it in a box. Total systems, theories of everything—when they are put on paper, experimented with, and acted upon in order to have control over things and people—are always a bit strange. This, in my view, was the project of human engineering, both its charm and its downfall.

Yes, charm. Despite the fact that the outcome of human and social engineering has often been dire, the undertaking had a quixotic appeal. No doubt *Brave New World* overtones are present in any project designed to run human society in a seamlessly efficient manner—with fears and pleasures doled out from each according to his ability, to each according to his malleability—but even Aldous Huxley could not have imagined the untoward form this project would take in late-twentieth- and early-twenty-first-century America. The reality has been less likely than any fiction. This book describes an experimental trajectory that, at its high point, not only tilted at windmills but tipped more than one over.

The story of human engineering started with Loeb's tropisms, Watson's rats-running-mazes, Ruml's social science, and Hawthorne's factory workers. A host of experimentalists used the laboratory (and other controlled spaces) to gather knowledge and build theories of human behavior with predictive value. They dreamed of a social and cultural matrix where people's actions and, eventually, thoughts could be engineered. They grappled with the age-old question of how the physical and the metaphysical are joined. In the end, their work influenced social patterns, individual activities, and inner selves. In the 1920s and 1930s using lab rats, the most malleable creatures available, allowed "rat researchers" to realize, chart, quantify, and publish every possible permutation of decision-making and trial-and-error behavior. The mazes and other devices these scientists built were like a Lilliputian Disney World for science. Their experiments were central to the largest and best-funded social science project in history, and their accomplishments included the intelligence test, the SAT, the opinion survey, the early poll, the projective test, the propaganda campaign, the anthropological databank, the "therapeutic situation" as a model for a new social authority, the focus group, and the most effective methods of coercive interrogation.

A key episode in the development of human engineering took place in 1938–41 when the core group of the Yale Institute of Human Relations merged behaviorist and Freudian science and tried to create a single unified system that would explain and predict not only outer physical but inner mental activities. They built their byzantine system on the backs of rats, punishment grills, and celluloid dolls, and they nurtured it in obscure academic redoubts. In a surprising twist, however, their work proved useful in many unanticipated ways to a mid-level segment of pragmatic social adjusters such as human relations personnel, CIA bureaucrats, personality-testers, military government officers, poll-takers, and opinion-makers. Meanwhile, scientists employed the Yale system of neobehaviorism in therapeutic situations (through the Dollard-Miller hypothesis), in running concentration camps (Poston Relocation Center for Japanese-Americans), and in governing occupied areas (Micronesia and Japan). It continues to form much of today's baseline, commonsense thinking in America—why people do what they do and how frustration is linked to aggression.

The successes and failures of human engineers are difficult to underestimate. Some critics found the work of Loeb, Watson, Hull, Mowrer, Dollard, Miller, and others ludicrous. And when the anthropologists who invented the Yale files went into war service and joined the Pacific campaign, the Yale institute was reduced to penury, forced to turn over its custom-built white stone hall to the medical school. By then, it seemed a monument to its own demise, and passersby could only wonder at the words engraved on the facade: INSTITUTE OF HUMAN RELATIONS.

But the exorbitance of human and social engineering produced successes, too. Their projects attracted untold excrescences of money: over $40 million in the 1920s to Ruml's Rockefeller beneficiaries in the interdisciplinary social sciences, about $20 million in the 1930s to the Yale institute, another million or so of wartime money to Yale's anthropological endeavors, and $25 million to MK-ULTRA alone, which was only one of the CIA's many programs and conduits for funding such research. After World War II, as scientists returned from war service, their insights and assumptions spread quickly and thoroughly throughout universities, think tanks, and policy outlets. (Almost 90 percent of American social scientists worked in some capacity for the government and military during the war.) An age of triumph for the "behavioral sciences" followed during the cold war, so much so that it would be nearly impossible to make a full account of all the projects in this vein—interdisciplinary, cross-cultural, social scientific—funded by U.S. government, military, and private sources. During the same years the Soviets tried their own experiments in social control, but they tended toward "hard" totalitarian techniques. Americans had a softer approach: to "operationalize" their theories combining Freud and behaviorist models of the inner self.

The biggest success has been the strangest of all. The odd obsessions of hermetic experimentalists produced an array of social science devices that flooded social life in secular twentieth- and twenty-first-century America. Suggestion-making apparatuses, mass-scale measurements, ethnographic marketing, opinion-watching and opinion-generating polls, psychological tests that tell you what you are like or what you will become, the fulmination of dreams and propagation of fears, all were ways to craft people's reactions to their social environment. And as people reacted to their surroundings, their responses in turn acted back on the environment, and the bombardment by messages, tools, and tests intensified, its focus sharpened. The result was the "scientific management of instinctual needs"[1]—a fulfillment of Thomas Carlyle's 1830s-era alarum, "Not the external and physical alone is now managed by machinery, but the internal and spiritual also."

Quixotism or charm aside, the human engineering movement with its experimental compartments and special devices caused much animal and human suffering. And many people continue to suffer from the use of these techniques outside the laboratory today, because they tend to magnify impulses of self-satisfaction, self-doubt, anxiety, and desire. For example, the consumer attitude survey encourages one to think that one's choice of a product among an array of products has real, important significance. The political opinion poll can skew results or entirely mislead (as the most recent election demonstrated). The focus group applied to movies makes a mockery of art; applied to product development, it caters to the dumb and transitory; applied to government policies, it leads to kowtowing. But even more than their techniques, human engineers' ideas have won out. The pervasiveness of frustration-aggression explanations for human behavior encourages a general tendency in media and society to think without thinking: it has become common to assume one has to "get out" one's aggression by exercising or expressing it, and not by understanding its nature or source. The therapeutics of human engineering militated against achieving any insight but the most banal. Human engineering's assumptions and methods are commonplace in many of the most unlovable aspects of modern living.

IN THE EXPERIMENTS DESCRIBED HERE, the scientists were also, in a sense, lab animals and human subjects. Although they believed in a firm separation between themselves and those who ran their labyrinthine mazes, sat in their problem boxes, were chained to their punishment grids, acted out grotesque parodies of human dramas, and went through endless rounds of reconditioning and rewiring, in fact such separation never existed. The experimenter cannot be distinguished from the experiment. Designed to be separate and perfectly measured, they were unavoidably connected. This led to a historical pivot sometime in the 1960s, a point at which the experimental momentum of human engineering became so great that it slipped into society at large and overflowed its bounds, no longer contained or containable in a laboratory.

At the same time, the "objects" of the experiments were also subjects, for one of the main results of human engineering was that each person in American society leads an experimental life. Each person conducts his or her own experiments in living, loving, re-creating, chemically altering, and dying. For example, the co-optation in the 1960s of the corporate slogan "Better Living Through Chemistry" emphasizes the way smiling young "radicals" on the street embraced the mottos of their elders but applied them inversely and obtusely. Instead of experimenting on others in objective laboratories to promote societywide comfort and stability, one experimented on oneself in subjective laboratories for societywide destabilization. The experimental life took shape as a rebellion but quickly became part of the status quo—always guided, of course, by an extremely intrusive and in fact ineluctably powerful surrounding environment or "milieu."

Here, then, is the crux of the matter, an immanent possibility coupled to an imminent slavery. The human engineering movement's pragmatic realism and its desire to strip away dead metaphysical assumptions were freeing, but the realities were not. Rulers and experts have tried throughout the centuries to make an inescapable machine for conversion or elicitation, for population control or population harnessing, to compel someone to go to war or to work the fields and factories, to force someone to bend to the will of another, to die for the king or to live for the state. But the work of human engineers showed that the only method that will bring about a *true* change of being, a moment at which there is a shift, is a cooperative one. (Scientists researching "forceful indoctrination" for the CIA discovered you could create a vegetable or a zombie or an unwilling ideological convert, but to have a fully sentient person act in precisely the way you wanted was a different matter.) There must be some degree of agreement, collusion, consent. There is a point at which one goes along to get along, then forgets there was ever a choice in the matter. When many people do this at the same time, you have a social movement of unparalleled and intricate dullness.

In this manner the stimulus-response model became a living tool. The laboratory research from 1900 to 1963 was a preview of the kind of reality people now inhabit that fits its subjects like a glove, or perhaps a harness, so comfortably tailored they rarely realize they are wearing it.

IT WAS AS IF RESEARCHERS were running their experiments on a stage set in a theater without an audience, with results that were both dull and startling. On the one hand, results confirmed the obvious—that a rat does not like to be electrically shocked, for example, and will change its behavior patterns to avoid it—but on the other, when findings were incorporated into a systematic series of equations and models, they allowed scientists to achieve a new level of control over experimental subjects. The discovery that if one made the shocks unpredictable, and thus created an environment of escalating stress, lab animals would become so uncomfortable that they actually experienced punishment as a relief, had further applications. It was an ad-

vance look at a stressful society in which people would seek small alle-
viations to dispel a looming but unknowable pain. Scientists connected the laboratory to the world in new ways. Building tiny
controlled versions of reality enabled the process and results to become part of reality. The microcosm became the macrocosm. The laboratory became the world and extended into it.

Thus the experiments described in this book *are* true: they showed
that all systems of truth-seeking are doomed to be bound by the
conditions in which they originate. For decades, scientists attempted
a science of conditioning that would free them from their own conditions. (It is almost a cliché that the most ambitious human engineers came from undistinguished social backgrounds to arrive at a
level of eminence unprecedented in their family history. This was the
case with Watson, Hull, Mowrer, Dollard, and West, among others.)
They tried to eliminate their own past conditioning while circumscrib-
ing the present conditions of others: lab rats, neurotics, housewives,
perverts, junkies, babies, double agents, automatons, Boy Scouts, pigeons. But the promise of liberating oneself by controlling others—by
building structures in tiny rooms—proved elusive. Their experiments
rebounded, like some karmic rubber band. In the final analysis, human
engineers did not make the world as it is today, they only helped, just
as most people do at one point or another. I admit I started out dislik-
ing the figure of the human engineer for his posturing, his claims
about doing hard science, and his cruel machinery for a noble cause—
he was not nice to mice—but I grew to like him. He did not want to be
bound by his past or cantilevered into his future, and if he was willing
to step on others' toes to achieve a feeling of freedom, his mistakes
stemmed not from malice but from ignorance about the relationships
that connect things and people. At first I did not want to see myself in
his efforts, and then later, I couldn't help it: isn't he a bit like you
and me?

INTRODUCTION

1. The phrases are from "Report of the Department of Social Relations for the Year 1956–7," Harvard Archives, HUF 801.4156.5, but the language is typical of much of the postwar behavioral sciences.

CHAPTER 1. STRANGE FRUITS AND VIRGIN BIRTHS

1. Francis Bacon, *New Atlantis and The Great Instauration*, ed. Jerry Weinberger (1627; reprinted Arlington Heights, Ill.: Harlan Davidson, 1989), 73–75.

2. Loeb's wife, Anne, later described his antipathy to dull, repetitive tasks as an antipathy to the "cut and dried." Quoted in Philip Pauly, *Controlling Life: Jacques Loeb and the Engineering Ideal in Biology* (Berkeley: University of California Press, 1990), 12. My account of Loeb's engineering approach is indebted to Pauly's excellent book.

3. This opinion was confirmed in obverse by a conservative colleague who called Loeb an "apostle of lawlessness." James and Loeb's critics are quoted in Pauly, *Controlling Life*, 25.

4. Jacques Loeb, *Dynamics of Living Matter* (1906; reprinted Chicago: University of Chicago Press, 1921), 120.

5. Ibid., 124, emphasis added.

6. Quoted in Pauly, *Controlling Life*, 39, 50.

7. Theodore Porter argues that Mach's work had a "buddhistic" side in that it explored the physical and philosophical implications of the fact that "the world never holds still." See "The Death of the Object: Fin-de-Siècle Philosophy of Physics," in Dorothy Ross, ed., *Modernist Impulses in the Human Sciences, 1870–1930* (Baltimore: Johns Hopkins University Press, 1994), 138.

8. Quoted in Pauly, *Controlling Life*, 51.

9. Loeb's descriptions of his work are culled from letters, papers, and statements, quoted in Pauly, *Controlling Life*, 79, 86, 86, 81, 84.

10. Ibid., 102.

11. Luther Burbank to R. S. Woodward, June 23, 1906, Carnegie Institution Archives.

12. William James, *A Pluralistic Universe* (Lincoln: University of Nebraska Press, 1996), 23.

CHAPTER 2. RUNNING THE MAZE

1. See Watson's autobiography, "John B. Watson," in Carl Murchison, ed., *A History of Psychology in Autobiography* (Worcester, Mass.: Clark University Press, 1936). Dewey's attitude toward Watson is discussed in Kerry W. Buckley, *Mechanical Man: John Broadus Watson and the Beginnings of Behaviorism* (New York: Guilford, 1989), 78–80.

2. Quoted in Robert Boakes, *From Darwin to Behaviorism: Psychology and the Minds of Animals* (New York: Cambridge University Press, 1984), 143. My discussion of the origins of the laboratory rat is indebted to Boakes's excellent account.

3. Philip J. Pauly, "Modernist Practice in American Biology," in Dorothy Ross, ed., *Modernist Impulses in the Human Sciences, 1870–1930* (Baltimore: Johns Hopkins University Press, 1994), 283.

4. John B. Watson, *Animal Education: An Experimental Study on the Psychical Development of the White Rat, Correlated with the Growth of Its Nervous System* (Chicago: University of Chicago Press, 1903), 9.

5. The historian of science Philip Pauly considers the experiment "carefully framed and rigorously designed" (as did most of Watson's contemporaries), but Robert Boakes, a practicing scientist writing about animal experimentation, suggests that the research was too crude to generate good results. See Pauly, "Modernist Practice in American Biology," 286, and Boakes, *From Darwin to Behaviorism*, 147. Watson's paper was given a short notice as "Kinaesthetic Sensations: Their Role in the Reactions of White Rats to the Hampton Court Maze," *Psychological Bulletin* 4 (1907): 211–12, and was published in full as "Kinaesthetic and Organic Sensations: Their Role in the Reactions of the White Rat to the Maze," *Psychological Review Monograph Supplements* 8, no. 2 (1907): 1–100.

6. *New York Times* editorial, January 1, 1907, 8.

7. Watson, "John B. Watson," in Murchison, *A History of Psychology in Autobiography*, 271. The following quotations from Watson are found on 271–75.

8. Report of the Trustee Committee, "Principles Governing the Memorial's Program in the Social Sciences," November 23, 1928, Rockefeller Archive Center, RF, R.G. 3, Series 910, Box 2, Folder 11.

9. See Howard Segal, *Technological Utopianism in American Culture* (Chicago: University of Chicago Press, 1985), 123.

10. Dorothy Ross, "Modernist Social Science in the Land of the New/Old," in Ross, *Modernist Impulses*, 187.

11. John B. Watson, "The Place of the Conditioned-Reflex in Psychology," *Psychological Review* 23, no. 2 (March 1916): 89–116.

12. B. F. Skinner, *The Behavior of Organisms: An Experimental Approach* (New York: D. Appleton-Century, 1938), 47.

13. Skinner queried "why the self behaves as it does" in *The Behavior of Organisms*, 3.

14. The speech was published as "Psychology as the Behaviorist Views It," *Psychological Review* 20 (March 1913): quotations from Watson's manifesto are from pages 158, 166–68, and 171. Recently the manifesto has been compared to Kandinsky's or the Futurists' work. Others emphasize that it advanced an "extreme behaviorism that . . . taught that all mental action could be ultimately explained as reflex responses to the environment." Dorothy Ross, *The Origins of American Social Science* (Cambridge: Cambridge University Press, 1991), 312.

15. The persistent health of behaviorism, despite periodic declarations of its death, can be seen in the alarms it has continued to set off. In 1951 the cul-

ture critic Marshall McLuhan blamed "behaviorist success manuals" for the war machine and the men who blindly staffed it, as well as an America in which people were generally encouraged to act as replaceable human parts. In 1963 the literary critic Joseph Wood Krutch characterized behaviorism as a meretricious science, the main function of which was to act as "calipers on the human mind." In 1972 the MIT linguist and social theorist Noam Chomsky made a thoroughgoing attack on behaviorism as a science, focusing especially on B. F. Skinner. In 1973 the psychologist Erich Fromm devoted a large part of a five-hundred-page book to discussing the continuing undeniable influence—and shortcomings—of behaviorism felt throughout the social sciences. See McLuhan, *The Mechanical Bride: Folklore of Industrial Man* (Boston: Beacon, 1951), 37; Krutch, "Calipers on the Human Mind," *Saturday Review*, June 19, 1965; Chomsky, "The Case Against B. F. Skinner," *New York Review of Books*, December 30, 1971.

16. See Benjamin Harris, "Whatever Happened to Little Albert?" *American Psychologist* 34, no. 2 (February 1979): 151–60. Although the article does not reveal what happened to Albert, it does make the point that the experiment has been very often misrepresented, that its claims for the effects of stimulus generalization have been exaggerated, and that Watson encouraged such exaggerations by suggesting Little Albert had become terrified of all white objects and all furry objects, which was not the case.

17. Watson, "John B. Watson," 281.

18. Watson quoted in Buckley, *Mechanical Man*, 137.

19. Quoted in Roland Marchand, *Advertising the American Dream: Making Way for Modernity 1920–1940* (Berkeley: University of California Press, 1985), 69.

20. Ann Hulbert, *Raising America: Experts, Parents, and a Century of Advice About Children* (New York: Alfred A. Knopf, 2003), 140.

21. John B. Watson, *Behaviorism* (1924; reprinted New York: W. W. Norton, 1970), x.

CHAPTER 3. EMBRACING THE REAL

1. Robert Hutchins quoted in Alva Johnson, "The National Idea Man," *New Yorker*, February 10, 1945, 8.

2. Assessment of Ruml by Martin Bulmer and Joan Bulmer, "Beardsley Ruml and the Laura Spelman Rockefeller Memorial," *Minerva* 19, no. 3 (1981): 347–407; Brownlow is quoted on 358.

3. The urge to "embrace the real," which fueled the growth of American social science during the twentieth century, depended on a relatively new attitude toward facts themselves, as the sociologist Nathan Glazer has argued: "From the beginning of the nineteenth century . . . a remarkable change in viewing man and society began to make itself felt. . . . What was truly new was the rise of a belief that the world, the social world around one, was not known, and that its reality could not be grasped by reading books, listening to learned men, or reflecting upon the facts of one's experience" (Nathan Glazer, "The Rise of Social Research in Europe," in Daniel Lerner, ed., *The Human Meaning of Social Sciences* [New York: Meridian, 1959], 46). The phrase "embrace of the real" is from Susan Sontag, *On Photography* (New York: Farrar, Straus and Giroux, 1977), 27.

4. Oliver Zunz, "Producers, Brokers and Users of Knowledge: The Institutional Matrix," in Dorothy Ross, ed., *Modernist Impulses in the Human Sciences, 1870–1930* (Baltimore: Johns Hopkins University Press, 1994), 293–94.

5. Quoted in Richard Gillespie, *Manufacturing Knowledge: A History of the Hawthorne Experiments* (Cambridge: Cambridge University Press, 1991), 32.

6. On the situation of the wealthy and the early years of foundations, I have relied on Judith Sealander, *Private Wealth and Public Life: Foundation Philanthropy and the Reshaping of American Social Policy from the Progressive Era to the New Deal* (Baltimore: Johns Hopkins University Press, 1997), 9ff.

7. Memoranda from the Rockefeller Foundation's Program and Policy Files: January 21, 1914, October 1922; "General Memorandum," December 11, 1933; memo from Edmund Day to Raymond Fosdick, Rockefeller Archive Center, RF, R.G. 3, Series 910, Box 2, Folders 10–13.

8. Quoted in Barry D. Karl and Stanley N. Katz, "Foundations and Ruling Class Elites," *Daedalus* 116 (1987): 13. Karl and Katz argue that the "managerial elite" of the big foundations did not intend to be servants of industrial capitalism. See also Donald Fisher, *Fundamental Development of the Social Sciences: Rockefeller Philanthropy and the United States Social Science Research Council* (Ann Arbor: University of Michigan Press, 1993), which focuses on the "exchange of power and resources between the academy, the economy, and the State."

9. Ron Chernow, *Titan: The Life of John D. Rockefeller, Sr.* (New York: Random House, 1990), 89. Big Bill wooed Rockefeller's mother (a well-brought-up woman of more substantial means than himself) while pretending to be a deaf-mute peddler, wearing a small slate around his neck that read "I am deaf and dumb." As Chernow tells it, "She was sufficiently taken in by his . . .

humbug that she involuntarily exclaimed in his presence, 'I'd marry that man if he were not deaf and dumb' " (7). And so she did.

10. The change in Spelman Memorial policy is discussed in General Memorandum of October 1922, Rockefeller Archive Center, RF, R.G. 3, Series 910, Box 2, Folder 10; Report of the Executive Committee and Director to the Board of Trustees for the year 1923–24, in "Memorial Policy in Social Science," Rockefeller Archive Center, RF, R.G. 3 Series 910, Box 2, Folder 10.

11. General Memorandum [by Beardsley Ruml], October 1922, Rockefeller Archive Center, RF, R.G. 3, Series 910, Box 2, Folder 10.

12. Ibid.

13. Report of the Trustee Committee, "Principles Governing the Memorial's Program in the Social Sciences," November 23, 1928, Rockefeller Archive Center, RF, R.G. 3, Series 910, Box 2, Folder 11. Emphasis added.

14. Except as otherwise noted, quotes are from Beardsley Ruml, "Recent Trends in Social Science," a presentation delivered at the dedication of the Social Science Research Building of the University of Chicago, December 17, 1929, RF, R.G. 3, Series 910, Box 2, Folder 12; emphasis added in the quote beginning "realistic contact . . ."

15. The phrase "grass-roots empiricism" is from the recollections of Don Martindale of the University of Minnesota, quoted in Robert C. Bannister, *Sociology and Scientism: The American Quest for Objectivity, 1880–1940* (Chapel Hill: University of North Carolina Press, 1987), 189–90.

16. However, this last, gigantic Rumlite project got its go-ahead just at the moment when the Spelman Memorial was absorbed into the larger Rockefeller armature and so was officially tabulated as the fruit of the foundation.

17. Dewey's letter to his wife is quoted in Robert Westbrook, *John Dewey and American Democracy* (Ithaca, N.Y.: Cornell University Press, 1991), 84.

18. Laski quoted in Bulmer and Bulmer, "Beardsley Ruml and the Laura Spelman Rockefeller Memorial," 400.

19. Quoted in Ann Hulbert, *Raising America: Experts, Parents, and a Century of Advice About Children* (New York: Alfred A. Knopf, 2003), 104.

20. Richard Gillespie, "The Hawthorne Experiments and the Politics of Experimentation," in Jill G. Morawski, ed., *The Rise of Experimentation in American Psychology* (New Haven, Conn.: Yale University Press, 1988), 101.

21. Ibid., 100–101.

22. Richard C. S. Trahair, "Elton Mayo and the Early Political Psychology of Harold D. Lasswell," *Political Psychology* (Fall/Winter 1981–82): 3, 176.

23. Quoted in Bulmer and Bulmer, "Beardsley Ruml and the Laura Spelman Rockefeller Memorial," 383.

24. Loren Baritz, *Servants of Power: A History of the Use of Social Science in American Industry* (Middletown, Conn.: Wesleyan University Press, 1960), chap. 5.

25. My account of the Hawthorne experiments is indebted to the excellent close study by Richard Gillespie, *Manufacturing Knowledge*.

26. Ibid., 4.

27. William H. Whyte, *The Organization Man* (1956; reprinted Philadelphia: University of Pennsylvania Press, 2002), 36.

28. Walter Lippmann, *Drift and Mastery* (1914; reprinted Madison: University of Wisconsin Press, 1985), 151.

CHAPTER 4. PSYCHIC MACHINES

1. John Dollard to president of Rockefeller Foundation, July 5, 1939, Rockefeller Archive Center, RF, R.G. 1.1, Series 200, Box 68, Folder 812.

2. Editorial, *New York Times*, February 16, 1929. A few months later, another headline described the institute's boldly meliorative aim: "To Seek Crime Correction," *New York Times*, April 19, 1929.

3. Angell quoted in Mark May, *Toward a Science of Human Behavior: A Survey of the Work of the Institute of Human Relations at Yale Through Two Decades, 1929–1949* (New Haven, Conn.: Yale University Press, 1950), 1.

4. Report of Miss Belcher, "Delinquency Studies," May 1933, Yale Archives, YRG 37-V, Series II, Box 6, Folder 18.

5. Allen Gregg, "Impressions from Visit to the Institute of Human Relations," October 7–12, 1935, Rockefeller Archive Center, RF, R.G. 1.1, Series 200, Box 67, Folder 808.

6. Hull memorandum, March 7, 1934, Rockefeller Archive Center, RF, R.G. 1.1, Series 200, Box 67, Folder 807.

7. May, *Toward a Science of Human Behavior*, 4. See also Mark May, "A Retrospective View of the Institute of Human Relations at Yale," *Behavior Science Notes* 3 (1971).

8. Edmund Day to Raymond Fosdick, July 28, 1936, Rockefeller Archive Center, RF, R.G. 1.1, Series 200, Box 67, Folder 809.

9. May, *Toward a Science of Human Behavior*, 7–13. The terms "pegs" and "holes" are May's.

10. Clark L. Hull, "Quantitative Aspects of the Evolution of Concepts," in *The Psychological Monographs* 28, no. 1 (1920): 1–85; from 1918 University of Wisconsin Ph.D. dissertation. Quotations in this section are from 3, 38.

11. Clark L. Hull, "The Concept of the Habit-Family Hierarchy and Maze Learning, Part I," *Psychological Review* 41 (1934): 40–41.

12. Clark L. Hull, "Goal Attraction and Directing Ideas Conceived as Habit Phenomena," *Psychological Review* 38 (1931): 502.

13. Laurence D. Smith, "Clark L. Hull: Background and Views of Science," *Behaviorism and Logical Positivism* (Stanford, Calif.: Stanford University Press, 1986), 149.

14. Stephen E. Toulmin and David Leary, "The Cult of Empiricism in Psychology and Beyond," in Sigmund Koch and David Leary, eds., *A Century of Psychology as Science* (Washington, D.C.: American Psychological Association, 1992), 606, 603.

15. John A. Mills, "The Genesis of Hull's *Principles of Behavior*," *Journal of the History of the Behavioral Sciences* 24 (1988): 393.

16. Karl E. Scheibe, "Metamorphoses in the Psychologist's Advantage," in Jill G. Morawski, ed., *The Rise of Experimentation in American Psychology* (New Haven, Conn.: Yale University Press, 1988), 63.

17. Hull quotations in this and the following paragraph are from Clark L. Hull, "Simple Trial-and-Error Learning: A Study in Psychological Theory," *Psychological Review* 37 (1930): 255–56.

18. Clark L. Hull, "Knowledge and Purpose as Habit Mechanisms," *Psychological Review* 37 (1930): 511.

19. The Rockefeller officer remarked that Hull appeared uninterested in anything save for his own theories, "which he finds so rigorous and mathematically logical." Syndor W. Walker, interviews at Institute of Human Relations, February 21–22, 1938, Rockefeller Archive Center, RF, R.G. 1.1, Series 200, Box 68, Folder 811.

20. Clark L. Hull, "The Goal-Gradient Hypothesis Applied to Some 'Field-Force' Problems in the Behavior of Young Children," *Psychological Review* 45, no. 4 (1938): 273.

21. Deborah Coon, "Standardizing the Subject: Experimental Psychology, Introspection, and the Quest for a Technoscientific Ideal," *Technology and Culture* 34 (1993): 760.

22. All Hull quotations in this paragraph are from Clark L. Hull, "Clark L. Hull," in Edwin G. Boring et al., eds., *A History of Psychology in Autobiography* (Worcester, Mass.: Clark University Press, 1952), 4: 145–50.

23. Ibid., 148–49.

24. As Hull noted in one of his idea books, quoted in Smith, "Clark L. Hull," 162n49.

25. The address is published as Clark L. Hull, "Mind, Mechanism, and Adaptive Behavior," *Psychological Review* 44, no. 1 (January 1937): 1–32; the recollection of Alphonse Chapanis is quoted in Smith, "Clark L. Hull," 167n49.

26. Hull, "Goal Attraction and Directing Ideas," 502.

27. Jean Matter Mandler and George Mandler, "The Diaspora of Experimental Psychology: The Gestaltists and Others," in Donald Fleming and Bernard Bailyn, eds., *The Intellectual Migration, Europe and America in 1930–1960* (Cambridge, Mass.: Harvard University Press, 1969), 374–75.

28. Bruner was attending his first APA meeting when Hull unveiled his thinking machine. He recalls being less than entranced by "a reductionism that I found fascinating and enragingly detestable." Jerome S. Bruner, "Jerome S. Bruner," in Gardner Lindzey, ed., *A History of Psychology in Autobiography* (San Francisco: W. H. Freeman & Co., 1980), 86.

29. Sigmund Freud, *Civilization and Its Discontents* (New York: W. W. Norton, 1961), 104.

CHAPTER 5. CIRCLE OF FEAR AND HOPE

1. John A. Mills, "The Genesis of Hull's *Principles of Behavior*," *Journal of the History of the Behavioral Sciences* 24 (1988): 396.

2. Mowrer's work on rotating pigeons, which he continued for two years at Yale while working also with humans and other animals, was published as O. Hobart Mowrer, "An Analysis of the Effects of Repeated Bodily Rotation, with Especial Reference to the Possible Impairment of Static Equilibrium," *Annals of Otology, Rhinology and Laryngology* 43 (1934): 367–87. The effect on balance when the eyelids of pigeons have been sewn shut since birth appears in O. Hobart Mowrer, " 'Maturation' vs. 'Learning' in the Develop-

ment of Vestibular and Optokinetic Nystagmus," *Journal of Genetic Psychology* 48, no. 2 (1936). The quotation beginning "protracted aftereffects of continuous bodily rotation" is found in O. Hobart Mowrer, "O. H. Mowrer," in Gardner Lindzey, ed., *A History of Psychology in Autobiography* (Englewood Cliffs, N.J.: Prentice Hall, 1974), 336.

3. Ibid., 333.

4. M. J. Bass and C. L. Hull, "Irradiation of Tactile Conditioned Reflex," *Journal of Comparative Psychology* 17 (1937): 50.

5. Quotations in this and the three following paragraphs are from O. Hobart Mowrer, "Preparatory Set (Expectancy): A Determinant in Motivation and Learning," *Psychological Review* 45 (1938); quotations in this and the two following paragraphs are from pages 67, 72, 76, 77n, and 83.

6. William James, "The Chicago School," *Psychological Bulletin* 1, no. 1 (January 15, 1904): 2.

7. Clark L. Hull, "Knowledge and Purpose as Habit Mechanisms," *Psychological Review* 37 (1930): 512–14 passim.

8. That is, to *know* the world is in some sense to simulate it; see the suggestive discussion in Jean-Pierre Dupuy, *The Mechanization of the Mind: On the Origins of Cognitive Science* (Princeton, N.J.: Princeton University Press, 2000), 96. Contrast this view with the more doctrinaire one of B. F. Skinner, who specified very strictly, "The individual organism simply reacts to its environment, rather than to some inner experience of that environment." Quoted in O. Hobart Mowrer, "The Behavior Therapies, with Special Reference to Modeling and Imitation," *American Journal of Psychotherapy* 20 (1966): 444.

9. Henderson, quoted in Stephen J. Cross and William R. Albury, "Walter B. Cannon, L. J. Henderson and the Organic Analogy," *Osiris*, 2nd ser., 3 (1987): 182.

10. Mowrer, "O. H. Mowrer," 338. Quotations in the two following paragraphs are from pages 350 and 351.

11. O. Hobart Mowrer, "A Stimulus Response Analysis of Anxiety and Its Role as Reinforcing Agent," *Psychological Review* 46 (1939): 558. Mowrer was not the only one in the field of aversive learning, but he was most certainly a leader: "Mowrer himself was one of the leading innovators in this methodological and conceptual evolution," Herrnstein commented in his overview of the field, "Method and Theory in the Study of Avoidance," *Psychological Review* 76 (1967): 53.

12. O. Hobart Mowrer, *The New Group Therapy* (Princeton, N.J.: Van Nostrand, 1964), 68, 85. Except as otherwise noted, all quotations in the following four paragraphs are from the same source, pages 8, 9, 39, 52, and 147.

13. Thomas Merton, *The New Man* (New York: Farrar, Straus and Giroux, 1961), 120.

14. O. Hobart Mowrer, "The Recovery of Responsibility," in *New Group Therapy*, 11; italics in original.

CHAPTER 6. IN AND OUT OF THE SOUTH

1. Simone de Beauvoir to Jean-Paul Sartre, February 24, 1947, in *Letters to Sartre* (New York: Arcade, 1993).

2. John Dollard, *Caste and Class in a Southern Town* (1937; reprinted New York: Doubleday, 1957), 242.

3. John Dollard, "Yale's Institute of Human Relations: What Was It?" *Ventures* (Winter 1964): 32.

4. John Dollard and Neal E. Miller, *Personality and Psychotherapy: An Analysis in Terms of Learning, Thinking, and Culture* (New York: McGraw-Hill, 1950), 448.

5. William Fielding Ogburn quoted in Walter A. Jackson, *Gunnar Myrdal and America's Conscience: Social Engineering and Racial Liberalism, 1938–1987* (Chapel Hill: University of North Carolina Press, 1990), 96.

6. Lawrence K. Frank memorandum, "Study of Comparative Cultures," Fall 1927, Rockefeller Archive Center, RF, R.G. 1.1, Series 200, Box 408, Folder 4828.

7. Edward Sapir and John Dollard, interview by Stacy May, March 29, 1933, Rockefeller Archive Center, RF, R.G. 1.1, Series 200, Box 408, Folder 4830.

8. Dollard, quoted in Steven Weiland, "Life History, Psychoanalysis, and Social Science: The Example of John Dollard," *South Atlantic Quarterly* 86, no. 3 (1987): 276.

9. In the discussion of Dollard's book that follows, quotations are from *Caste and Class*, pages 3, 8, 15, 20, 34, 35–36, 48, 61, 129, 174, 180, 242–44, 329, 333, 335, and 350.

10. Grace Elizabeth Hale, *Making Whiteness: The Culture of Segregation in the South, 1890–1940* (New York: Pantheon, 1998), 17.

11. W.E.B. Du Bois, review of Hortense Powdermaker's *After Freedom*, in *Social Forces* 18, no. 1 (October 1949): 138. Emphasis added.

12. John Dollard to Margaret Mead, July 22, 1935, Margaret Mead Papers, Library of Congress.

13. John Dollard to Margaret Mead, June 22, 1935, Margaret Mead Papers, Library of Congress.

14. C. Vann Woodward, *The Strange Career of Jim Crow* (London: Oxford University Press, 1966), 103–104.

15. Lyle H. Lanier, "Mr. Dollard and the Scientific Method," *Southern Review* 3 (1938): 660; emphasis in original.

16. John Dollard to Margaret Mead, October 8, 1937, Margaret Mead Papers, Library of Congress.

17. John Dollard to Edmund Day attaching 1934 Proposal, February 28, 1935, Rockefeller Archive Center, RF, R.G. 1.1, Series 200, Box 67, Folder 808.

18. See John Dollard, "Hostility and Fear in Social Life," *Social Forces* 17, no. 1 (1936).

19. Neal Miller, "The Value of Behavioral Research on Animals," *American Psychologist* 40, no. 4 (1985): 426, emphasis added.

20. Dollard and Miller, *Personality and Psychotherapy*, 447.

21. Dollard, "Yale's Institute of Human Relations," 38.

22. According to Sydnor Walker's evaluation of February 24, 1938, Miller was "studying conflict situations on rats as a preface to studying conflict situations on humans. . . . M. strikes me as one of the psychologists who are interested in the subject although they have no natural endowment in that direction." Rockefeller Archive Center, RF, R.G. 1.1, Series 200, Box 68, Folder 811.

23. Rockefeller and Yale records further noted with some relief, "Group at Yale are working much more effectively and concentratedly now than previously. Dollard himself has settled down and is rather less extremist in statement and point of view than he was." Allen Gregg diary re. lunch with Dollard, December 9, 1939, Rockefeller Archive Center, RF, R.G. 1.1, Series 200, Box 68, Folder 812. The "consistently represented" point is made in Leonard Doob to Mark May, October 4, 1937, Yale Archives, YRG 37-V, Series II, Box 7, Folder 40.

CHAPTER 7. AN ORDINARY EVENING IN NEW HAVEN

1. The experiments were published as H. D. Lasswell, "Verbal References and Physiological Changes During the Psychoanalytic Interview," *Psychoanalytic Review* 22 (1935), and H. D. Lasswell, "Certain Prognostic Changes During Trial (Psychoanalytic) Interviews," *Psychoanalytic Review* 23 (1936). Lasswell does not specify the actual dates of his experiments. They were run in two batches, with two different groups of patient-volunteers. The earliest experiment was apparently in 1927, while Lasswell was spending six months studying with Elton Mayo at the Harvard Business School. See also R. Trahair, "Elton Mayo and the Early Political Psychology of Lasswell," *Political Psychology* 3 (1981–82), and Roy R. Grinker, Sr., "Psychoanalysis and Autonomic Behavior," in Arnold A. Rogow, ed., *Politics, Personality, and Social Science in the Twentieth Century: Essays in Honor of Harold D. Lasswell* (Chicago: University of Chicago Press, 1969).

2. Marie Jahoda, "The Migration of Psychoanalysis: Its Impact on American Psychology," in Donald Fleming and Bernard Bailyn, eds., *The Intellectual Migration: Europe and America, 1930–1960* (Cambridge, Mass.: Harvard University Press, 1969), 426.

3. Clark L. Hull, "A Primary Social Science Law," appendix to Mark May, *Toward a Science of Human Behavior: A Survey of the Work of the Institute of Human Relations Through Two Decades, 1929–1949* (New Haven, Conn.: Yale University Press, 1950), 85, emphasis in original.

4. Neal R. Miller, "Theory and Experiment Relating Psychoanalytic Displacement to Stimulus-Response Generalization," *Journal of Abnormal and Social Psychology* 43, no. 2 (1948): 156; the paper was originally presented in September 1939.

5. John Demos, "Oedipus and America: Historical Perspectives on the Reception of Psychoanalysis in the United States (1978)," in Joel Pfister and Nancy Schnog, eds., *Inventing the Psychological: Toward a Cultural History of Emotional Life in America* (New Haven, Conn.: Yale University Press, 1997).

6. Nathan Hale, Jr., *Freud and the Americans: The Beginnings of Psychoanalysis in the United States, 1876–1917* (New York: Oxford University Press, 1971), 363.

7. See Eli Zaretsky, *Secrets of the Soul: A Social and Cultural History of Psychoanalysis* (New York: Alfred A. Knopf, 2004): 147, and Christopher Lasch, *The New Radicalism in America, 1889–1963* (New York: W. W. Norton, 1965), 105–107. Luhan's "Notes upon Awareness" was dated 1938.

8. Joel Pfister, "On Conceptualizing the Cultural History of Emotional and Psychological Life in America," in Pfister and Schnog, *Inventing the Psychological*, 28.

9. Clark L. Hull, "Simple Trial-and-Error Learning: A Study in Psychological Theory," *Psychological Review* 37 (1930): 250.

10. Robert Sears, "Psychoanalysis and Behavior Theory, 1907–1965," in Sigmund Koch and David Leary, eds., *A Century of Psychology as Science* (Washington, D.C.: American Psychological Association, 1992), 214.

11. Clark L. Hull memorandum, November 17, 1936, quoted in David Shakow and David Rapaport, "The Influence of Freud on American Psychology," *Psychological Issues* 4, no. 1 (1964): 140.

12. O. Hobart Mowrer, "O. H. Mowrer," in Gardner Lindzey, ed., *A History of Psychology in Autobiography* (Englewood Cliffs, N.J.: Prentice Hall, 1974), 337.

13. On the triumph of the frustration-aggression hypothesis, see Erich Fromm, *The Anatomy of Human Destructiveness* (New York: Holt, 1973), 91; Gregg Mittman "Dominance, Leadership, and Aggression: Animal Behavior Studies During World War II," *Journal of the History of the Behavioral Sciences* 26 (1990): 11; and John Dower, *Embracing Defeat: Japan in the Wake of World War II* (New York: W. W. Norton, 1997), 214.

14. Comments on the papers of the frustration-aggression unveiling, as well as the judgment that the hypothesis was essentially inconsistent with Freud, are from Shakow and Rapaport, "Influence of Freud on American Psychology," 141ff.

15. "The earlier approaches [to Freud] via integration, translation, and verification gave way to absorption. Behavior theory absorbed the subject matter, concepts, and principles from psychoanalysis but ignored the analytic method, theoretical structure, and operations that served to define the concepts." Sears, "Psychoanalysis and Behavior Theory," 215.

16. Ellen Herman, *The Romance of American Psychology: Political Culture in the Age of Experts* (Berkeley: University of California Press, 1995), 36–38.

17. Shakow and Rapaport, "Influence of Freud on American Psychology," 141–42.

18. The drive-cue-response-reward variant of the Miller/Dollard formula appears in Mark May, *Toward a Science of Human Behavior: A Survey of the Work of the Institute of Human Relations Through Two Decades, 1929–1949* (New Haven, Conn.: Yale University Press, 1950), 12; originally from Miller

and Dollard's *Social Learning and Imitation* (New Haven, Conn.: Yale University Press, 1941). Miller and Dollard pioneered the use of the "analytic situation" as a mini-laboratory for human engineering and concluded, "Culture, as conceived by social scientists, is a statement of the design of the human maze," *Social Learning and Imitation*, 5.

19. Quotations in this paragraph are from Dollard et al., *Frustration and Aggression* (New Haven, Conn.: Yale University Press, 1939), 58, 57, 72, 75, 122.

20. Untitled item under "People," *Time*, March 6, 1939.

21. James Baldwin, *Notes of a Native Son* (Boston: Beacon Press, 1950), 20.

22. Leonard Doob and Robert Sears, "Factors Determining Substitute Behavior and the Overt Expression of Aggression," *Journal of Abnormal and Social Psychology* 34, no. 3 (1939): 294, 298.

23. Anxiety can be created in this manner within the laboratory, as experiments with three-month-old infants proved; see Mowrer, "The Freudian Theories of Anxiety: A Reconciliation," Monday Night Group talk, 1941, Yale Archives, YRG 57-V, Boxes 28–31.

24. According to a 1977 review of the scientific literature by Fisher and Greenberg cited in Gail Hornstein, "The Return of the Repressed: Psychology's Problematic Relations with Psychoanalysis," *American Psychologist* 47, no. 2 (1992): 258n8.

25. Nikolas Rose, *Governing the Soul: The Shaping of Private Life* (London: Routledge, 1990); T. J. Jackson Lears, *Fables of Abundance: A Cultural History of Advertising in America* (New York: Basic Books, 1994), 138.

CHAPTER 8. THE BIGGEST FILE

1. C. Wright Mills wrote, "What must be grasped is the picture of society as a great salesroom, an enormous file, an incorporated brain, a new universe of management and manipulation." C. Wright Mills, *White Collar: The American Middle Class* (1951; reprinted New York: Oxford University Press, 1971), xv. The descriptions of the files' purpose are from Clellan S. Ford, "The Development of the *Outline of Cultural Materials*," *Behavioral Science Notes* 3 (1971): 176, and George P. Murdock, *Social Structure* (London: Macmillan, 1949), viii.

2. Quotations in this paragraph are from George P. Murdock, "The Cross-Cultural Survey," *American Sociological Review* 5 (1940): 362, and a George P. Murdock memorandum, "Proposed Program for Anthropological Re-

search under the Direction of the IHR as Part of a Coordinated Program of Research Aimed at the Achievement of an Integrated Social Science," c. August 1939, Yale Archives, YRG 37-V, IHR, Series II, Box 11, Folders 11–95.

3. Murdock invoking Spencer's vision, August 29, 1942, meeting of the advisory board of the Strategic Index of Latin America, National Archives, State Department, R.G. 229, Entry 1, Stack Area 350, Row 76, Compartment 2, Shelf 4, Box 134.

4. Mills, *White Collar*, xii.

5. Alexander Spoehr, "George Peter Murdock 1897–1985," *Ethnology* 24 (October 1985): 309.

6. See David Price, *Threatening Anthropology: McCarthyism and the FBI's Surveillance of Activist Anthropologists* (Durham, N.C.: Duke University Press, 2004).

7. George Peter Murdock, "Autobiographical Sketch," *Culture and Society* (Pittsburgh: University of Pittsburgh Press, 1965), 353.

8. Murdock, quoted in Mark May letter to Murdock, January 20, 1931, Yale Archives, YRG 37-V, IHR, Series II, Box 11, Folders 11–95.

9. Clellan S. Ford, "The Development of the *Outline of Cultural Materials*," 173–85.

10. *HRAF Research Guide* (New Haven, Conn.: Human Relations Area Files, 1965), 27.

11. Ford, "Development of the *Outline of Cultural Materials*," 8.

12. Mauss quoted in James Clifford, "On Ethnographic Surrealism," in *The Predicament of Culture: Twentieth-Century Ethnography, Literature and Art* (Cambridge, Mass.: Harvard University Press, 1988), 129; Clifford's comment is from page 127.

13. John Dollard and George P. Murdock, "Memorandum to Dr. Mark A. May Concerning a Research Job That Needs to Be Done," c. April 1936, Yale Archives, YRG 37-V, IHR, Series II, Box 11.

14. "The Cross-Cultural Survey," introduction to "Strategic Bulletins of Oceania," nos. 1–8, compiled by the Cross-Cultural Survey IHR, Yale University, Restricted. National Archives: Office of Naval Intelligence monograph file, Marshalls, Boxes 61–68, R.G. 38, Stack Area 370, Row 15, Compartment 25, Shelves 4–5.

15. See "Yale Joins Up," *Newsweek*, December 21, 1942, 76–77, and Yale News Bureau announcement quoted in "Yale University Integrates Its Entire Curriculum with War Effort," *School and Society* 55 (March 28, 1942): 357.

16. Robin Winks, *Cloak and Gown: Scholars in the Secret War, 1939–1961* (New Haven, Conn.: Yale University Press, 1997), 31.

17. Barry Katz, *Foreign Intelligence: Research and Analysis in the Office of Strategic Services, 1942–1945* (Cambridge, Mass.: Harvard University Press, 1989), 17.

18. Robert T. Miller to George Dudly, March 30, 1942, National Archives, State Department, R.G. 229, Entry 1, Stack Area 350, Row 76, Compartment 2, Shelf 4, Box 134: General Records, Central Files, Commercial and Financial Research.

19. Quoted in Katz, *Foreign Intelligence*, 18.

20. Proposal to Establish a Strategic Index for the Other American Republics (emphasis in original), March 30, 1942, and Project Authorization for Strategic Index of the Other American Republics, May 14, 1942, authorized by Directive Council and Psychological Warfare Division of Office of Inter-American Affairs, National Archives, State Department, R.G. 229, Entry 1, Stack Area 350, Row 76, Compartment 2, Shelf 4, Box 134.

21. George P. Murdock to Mark May, July 29, 1945; September 3, 1945; and September 13, 1945, Yale Archives, YRG 37-V, Series II, Box 11.

22. Clellan S. Ford, "The Role of HRAF in the Organization of Knowledge About Behavior and Mankind," *Behavior Science Notes* 1 (1966): 4.

23. Clellan S. Ford, "HRAF: 1949–1969, A Twenty-Year Report," *Behavior Science Notes* 5 (1970): 13.

24. Quoted in Ira Bashkow, "The Dynamics of Rapport in a Colonial Situation: David Schneider's Fieldwork on the Islands of Yap," in George W. Stocking, Jr., ed., *Colonial Situations: Essays on the Contextualization of Ethnographic Knowledge* (Madison: University of Wisconsin Press, 1991), 188.

25. Clellan S. Ford and Frank A. Beach, *Patterns of Sexual Behavior* (New York: Harper and Row, 1951), 2–7 passim. Quotations in the following three paragraphs are from pages 13, 15, and 21.

26. E. R. Leach, review of *Social Structure, Man* 50 (August 1950): 107–108.

27. Murdock quotations here and in the following paragraphs are from *Social Structure*, 2, 23, 30, 79–82, 178–83, 197, and 321.

28. Leach, review of *Social Structure*, 108, and Morris Edward Opler, review of *Social Structure, American Anthropologist* 52 (1950): 80.

29. John W. M. Whiting and Irvin L. Child, *Child Training and Personality: A Cross-Cultural Study* (New Haven, Conn.: Yale University Press, 1953), 63.

The following quotations in this section are from pages 61 (emphasis added), 62, 66, 67, 70, 94, and 103.

30. Ford, "Development of the *Outline of Cultural Materials*," 176.

CHAPTER 9. ANTHROPOLOGY'S LABORATORY

1. Official names for the islands have in most cases changed from those common during World War II, reflecting more accurately local usage and proper pronunciation. Truk is Chuuk, Kusaie is Kosrae, Kapingamarangai is Kapingamarangi, Ponape is Pohnpei. Here, when speaking of the war era, the older names will be retained.

2. Minutes of Meeting of Committee on Anthropology of Oceania, December 4, 1942, National Academy of Sciences–National Research Council Archives: Div. A&P: DNRC: A&P: Com. on Anthropology of Oceania: General: 1942–1943; 16.

3. "Navy Program for Occupied Areas," n.d., National Archives, Records of ONI, R.G. 38, Stack Area 370, Row 12, Compartment 9, Shelf 3, Box 51.

4. Press release (draft) from Naval Operations, February 19, 1944, National Archives, R.G. 38, Stack Area 370, Row 12, Compartment 9, Shelf 1, Boxes 32–33. Other training schools in military and civil affairs were run by the army at the University of Chicago, Stanford, and Harvard between December 1943 and October 1945; in addition, a few graduates of the U.S. schools attended courses at the British Civil Affairs Staff Center at Wimbledon, England.

5. Dorothy Richard, *United States Naval Administration of the Trust Territory of the Pacific Islands*, vol. 1 (Office of the Chief of Naval Operations, 1957), 50.

6. Arjun Appadurai, *Modernity at Large: Cultural Dimensions of Globalization* (St. Paul: University of Minnesota, 1994), 2.

7. H. L. Pence, 5th Draft of memorandum, May 5, 1944, R.G. 38, Stack Area 370, Row 12, Compartment 9, Shelf 3, Boxes 50–51.

8. Civil Affairs pamphlet, "What Is Civil Affairs?" May 3, 1944, Office of Chief of Naval Operations, Office of Assistant Chief, Island Governments Division, 1944, R.G. 38, Stack Area 370, Row 12, Compartment 9, Shelf 1, Boxes 32–33.

9. In December 1943 the *United States Army and Navy Manual of Military Government and Civil Affairs* set forth objectives guiding the armed forces of the United States for establishing military government and civil affairs con-

trol. The manual instructed: "Protect military personnel by care of the health and well-being of the civilian population."

10. Knowles A. Ryerson to Chester W. Nimitz, report, March 30, 1944, National Archives, R.G. 38, Stack Area 370, Row 12, Compartment 9, Shelf 1, Boxes 32–33.

11. Philip C. Jessup to H. L. Pence, March 16, 1944, National Archives, R.G. 38, Stack Area 370, Row 12, Compartment 9, Shelf 3, Box 49.

12. Leighton quoted in H. G. Barnett, *Anthropology in Administration* (Evanston, Ill.: Row, Peterson & Co., 1956), 54.

13. Alexander Leighton, *The Governing of Men: General Principles and Recommendations Based on Experience at a Japanese Relocation Camp* (Princeton, N.J.: Princeton University Press, 1946), 373.

14. Ryerson to Nimitz, report, March 30, 1944.

15. "Secret" document, April 14, 1944, ONI Monograph File, POA, Marshalls, R.G. 38, Stack Area 370, Row 15, Compartment 25, Shelves 4–5, Boxes 61–68.

16. Ira Bashkow, "The Dynamics of Rapport in a Colonial Situation: David Schneider's Fieldwork on the Islands of Yap," in George W. Stocking, Jr., ed., *Colonial Situations: Essays on the Contextualization of Ethnographic Knowledge* (Madison: University of Wisconsin Press, 1991), 180.

17. See his memo, April 10, 1944, National Archives, R.G. 38, Stack Area 370, Row 12, Compartment 9, Shelf 1, Boxes 32–33.

18. Bulletin re CIMA Project, May 13, 1947, National Academy of Science–National Research Council (hereinafter NAS–NRC) Archives, ADM, Ex. Bd., Pacific Science Board, CIMA.

19. George P. Murdock to Walter Miles, February 5, 1946, NAS–NRC Archives, Div. A&P: DNRC: A&P: Comm. on Anthropology of Oceania, General, 1942–3.

20. George P. Murdock, interview by Charles Dollard, December 16, 1945, Carnegie Corporation Archives, Cross-Cultural Survey, Box 377. Charles was John Dollard's brother.

21. Chester W. Nimitz to Dr. Ross Harrison, late 1946 or early 1947, NAS–NRC Archives, ADM, Ex. Bd., Pacific Science Board, CIMA.

22. P. F. Lee to Dr. Ross Harrison, December 24, 1946, NAS–NRC Archives, ADM, Ex. Bd., Pacific Science Board, CIMA.

23. Proposal for a Coordinated Investigation of the Micronesian Peoples (CIMP), c. Jan. 1947, NAS-NRC Archives, ADM, Ex. Bd., Pacific Science Board, CIMA.

24. Commander of the Marianas to navy personnel, memorandum, April 30, 1947, NAS–NRC Archives, ADM, Ex. Bd., Pacific Science Board, CIMA, General, 1947, and Bulletin from Pacific Science Office to CIMA Project Participants re CIMA Policy, August 5, 1947, NAS–NRC Archives, ADM, Ex. Bd., Pacific Science Board, CIMA.

25. Bashkow, "Dynamics of Rapport in a Colonial Situation," 222. All quotations in this excerpt are from Schneider's typed field notes; references have been deleted.

26. Quoted in Stewart Firth, *Nuclear Playground: Fight for an Independent and Nuclear Free Pacific* (Honolulu: University of Hawai'i Press, 1987), 42. On the trials of Bikini evacuees, see Robert C. Kiste, *Kili Island: A Study of the Relocation of the Ex-Bikini Marshallese* (Eugene: University of Oregon Department of Anthropology, 1968).

27. Paul Tibbets, film footage c. 1948, in the documentary film *The Atomic Cafe* (1982; Jayne Loader, director).

28. Ward Goodenough, "Anthropology in the Twentieth Century and Beyond," *American Anthropologist* 104 (June 2, 2002): 424.

29. J. E. Weckler, letter regarding censorship of film by Bentzen, October 4, 1951, NAS–NRC Archives, ADM, Ex. Bd., Pacific Science Board, CIMA.

30. Thomas Gladwin and Seymour Sarasen, *Truk: Man in Paradise* (New York: Viking Fund Publications in Anthropology, 1953). The lead author later came to question whether psychological tests do in fact measure what they claim to be measuring, eventually declaring himself a "nonanthropologist" to work for Micronesian rights. See Thomas Gladwin, *East Is a Big Bird* (Cambridge, Mass.: Harvard University Press, 1970), and William A. Lessa, *Ulithi: A Design for Living* (New York: Holt, Rinehart, and Winston, 1966), 12.

31. William A. Lessa and Marvin Spiegelman, *Ulithian Personality as Seen Through Ethnological Materials and Thematic Test Analysis*, University of California Publications in Culture and Society 2, no. 5 (Berkeley: University of California Press, 1954), 279–86.

32. Melford E. Spiro, "A Psychotic Personality in the South Seas," NAS–NRC Archives, ADM, Ex. Bd., Pacific Science Board, CIMA. Later published as Melford Spiro, "A Psychotic Personality in the South Seas," *Psychiatry* 13, no. 2 (1950): 189–204.

33. D. W. Brogan, *The American Character* (New York: Vintage, 1956), 199.

CHAPTER 10. THE IMPOSSIBLE EXPERIMENT

1. Shana Alexander, *Anyone's Daughter: The Times and Trials of Patty Hearst* (New York: Viking, 1979), 506.

2. Ibid., 508.

3. Quoted in Martin A. Lee and Bruce Shlain, *Acid Dreams: The Complete Social History of LSD: The CIA, the Sixties, and Beyond* (New York: Grove Press, 1985), 27.

4. A CIA internal memorandum from 1963 discussing in retrospect the behavior-control programs the agency had run, quoted in Patricia Greenfield, "CIA's Behavior Caper," *APA Monitor*, December 1977, 1.

5. Louis Jolyon West, "The Future of Psychiatric Education," *American Journal of Psychiatry* 130, no. 5 (May 1973). West's vision of a future unified science of behavior is found on 522–26.

6. Quoted in Alexander, *Anyone's Daughter*, 515.

7. Harry Kreisler, "Evil, the Self, and Survival: Conversation with Robert Jay Lifton, M.D.," November 2, 1999; online at *http://globetrotter.berkeley.edu/people/Lifton/lifton-con2.html*.

8. Eugene Kinkead, "A Reporter at Large: A Study of Something New in History," *New Yorker*, October 26, 1957, 102–53. Ultimately not many of these cases went to court martial, and only eleven servicemen were convicted of treasonous collaboration with the enemy.

9. Robert J. Lifton, "Home by Ship: Reaction Patterns of American Prisoners of War Repatriated from North Korea," *American Journal of Psychiatry* 110 (1954): 734–35.

10. Raymond Bauer and Edgar Schein, introduction to special "Brainwashing" issue, *Journal of Social Issues* 13, no. 3 (1957): 4.

11. Quoted in Kinkead, "Reporter at Large," 118.

12. I. E. Farber, Harry Harlow, and Louis Jolyon West, "Brainwashing, Conditioning, and DDD (Debility, Dependency, Dread)," *Sociometry* 20 (1957): 282; quotations from this article in the following two paragraphs are from pages 275 and 278.

13. "Forceful indoctrination" was a preferred term, superior to "brainwashing" because less incendiary. See Louis Jolyon West, "United States Air Force Prisoners of the Chinese Communists," in Group for the Advancement of Psychiatry, *Methods of Forceful Indoctrination: Observations and Interviews*

(Group for the Advancement of Psychiatry Publications, 1957), symposium 4.

14. West, "Future of Psychiatric Education," 525.

15. Sidney Cohen, *The Beyond Within: The LSD Story* (New York: Atheneum, 1964). Animal and human research with LSD is summarized on 34–37ff. Dr. Cohen also received CIA money for his research.

16. Louis Jolyon West et al., "Lysergic Acid Diethylamide: Its Effect on a Male Asiatic Elephant," *Science* 138 (December 7, 1962): 1100–102. West's comment about miscalculating the dosage comes from an interview in Alexander, *Anyone's Daughter*, 275.

17. Louis Jolyon West, "Flight from Violence: Hippies and the Green Rebellion," *American Journal of Psychiatry* 125, no. 3 (1968): 365; see also L. J. West and J. R. Allen, "Three Rebellions: Red, Black and Green," in J. H. Masserman, ed., *Science and Psychoanalysis* (New York: Grune and Stratton, 1968), vol. 13.

18. Jill G. Morawski, "Impossible Experiments and Practical Constructions: The Social Bases of Psychologists' Work," in Jill G. Morawski, ed., *The Rise of Experimentation in American Psychology* (New Haven, Conn.: Yale University Press, 1988), 73, 80. "Sensitive research programs" are described in a 1964 memo to the director of Central Intelligence, cited in John Ranelagh, *The Agency: The Rise and Decline of the CIA* (New York: Simon and Schuster, 1987), 205n35.

19. Thomas Almy, "Harold George Wolff, 1898–1962," *Transcripts of the Association of American Physicians* 75 (1962): 45–47.

20. Wolff quoted in John Marks, *The Search for the Manchurian Candidate: The CIA and Mind Control* (New York: W. W. Norton, 1991), 157–58.

21. Interview with Hinkle quoted in Marks, *Search for the Manchurian Candidate*, 136.

22. Lawrence E. Hinkle and Harold G. Wolff, "Communist Interrogation and Indoctrination of 'Enemies of the State,'" *Archives of Neurology and Psychiatry* 76 (1956): 115–74; the quotation that follows is from page 171.

23. Quoted in Marks, *Search for the Manchurian Candidate*, 139.

24. In fact, Wolff's theory was an autobiographical description of his own crippling headaches; as more recent researchers have shown, such personality dynamics need not lead to migraine, and not all migraines occur in people who fit the "driven" personality type that Wolff described. As one prominent researcher has remarked, "When you read Wolff's book you are reading

about Wolff." Dr. Arnold Friedman is quoted in J. N. Blau, "Harold G. Wolff: The Man and His Migraine," *Cephalalgia* 24 (2004): 217. See also Helen Goodell, "Thirty Years of Headache Research in the Laboratory of the late Dr. Harold G. Wolff," *Headache* 6 (January 1967).

25. Wolff quoted in Marks, *Search for the Manchurian Candidate*, 157.

26. Shulman interview quoted in Greenfield, "CIA's Behavior Caper," 6.

27. Quoted in Marks, *Search for the Manchurian Candidate*, 161.

28. CIA document, February 7, 1952, quoted in Lee and Shlain, *Acid Dreams*, 8–9n.

29. Rockefeller records quoted in Marks, *Search for the Manchurian Candidate*, 141.

30. D. Ewen Cameron, "Psychic Driving," *American Journal of Psychiatry* 112 (1956): 503. The comparison of interrogator to psychiatrist is from page 508. The description in this paragraph of the experiment's design is from pages 502–509.

31. The Society for the Investigation of Human Ecology funded Cameron's research from 1957 to 1960, totaling about $62,000. A far larger amount of around half a million came from the Canadian government's National Department of Health and Welfare; even after the CIA defunded Cameron's program, this source continued with $51,000. The Cameron program appears in the CIA's files as MK-ULTRA Subproject #68, including in the archives a document "Application for Grant to Study the Effects upon Human Behavior of the Repetition of Verbal Signals," dated January 21, 1957. Note that Cameron could have gotten funding elsewhere quite easily, as prestigious as he and his hospital were. See Marks, *Search for the Manchurian Candidate*, 140–48.

32. D. Ewen Cameron, "Production of Differential Amnesia as a Factor in the Treatment of Schizophrenia," *Comprehensive Psychiatry* 1 (1960): 27.

33. Linda MacDonald (Cameron patient) interview in documentary film *Mind Control* (1997, Turner Entertainment).

34. Donald Hebb quoted in Marks, *Search for the Manchurian Candidate*, 146.

35. Interviews in this paragraph with Orne, Schein, Shulman, Hinkle, and Hall (as well as others) are included in Greenfield, "CIA's Behavior Caper"; the CIA operative is quoted in Marks, *Search for the Manchurian Candidate*, 189.

36. Quoted in Marks, *Search for the Manchurian Candidate*, 154.

37. Lesley Gill, *The School of the Americas* (Durham, N.C.: Duke University Press, 2004), 65.

38. Ellen Herman, *The Romance of American Psychology: Political Culture in the Age of Experts* (Berkeley: University of California Press, 1955), 155–56. Military planners at a 1962 symposium of over three hundred social and behavioral scientists noted, "The kind of underlying knowledge required is the *understanding and prediction of human behavior at the individual, political and social group, and society levels*" (150). See also Oscar Salemink, "*Mois and Maquis*: The Invention and Appropriation of Vietnam's Montagnards from Sabatier to the CIA," in George W. Stocking, Jr., ed., *Colonial Situations* (Madison: University of Wisconsin Press, 1995), 274.

CHAPTER 11. THE REAL WORLD

1. Milgram eventually tried many variations on the basic experiment. In the two best-known experiments, when the experimenter was standing near the subject and the victim was in another room (heard but not seen), the results of obedience rates were 63 percent at Yale and 48 percent at a less imposing location fictitiously named "Research Associates of Bridgeport." In eighteen other trials with varying conditions, compliance ranged from 93 percent (when the participant did not have to administer shocks personally) to 0 percent (when two authorities gave contradictory orders, when the experimenter was the victim, and when the victim demanded to be shocked). See Stanley Milgram, *Obedience to Authority: An Experimental View* (New York: Harper and Row, 1974), tables 2, 3, 4, and 5.

2. On Milgram's background, see Ian Parker, "Obedience," *Granta* 71 (Autumn 2000). An ABC made-for-television movie, *The Tenth Level*, and a London play, *The Dogs of Pavlov*, were both based on the experiments and their implications.

3. Stanley Milgram, "Behavioral Study of Obedience," *Journal of Abnormal and Social Psychology* 67 (1963): 371–78; Milgram, "Some Conditions of Obedience and Disobedience to Authority," *Human Relations* 18 (1961): 57–75.

4. Hannah Arendt, *Eichmann in Jerusalem: A Report on the Banality of Evil* (New York: Penguin, 1992). Quotations in this paragraph and the next are from pages 25 and 95.

5. Milgram, "Behavioral Study of Obedience," 375.

6. Louis Menand, "Why Do People Follow Dictators?" *New Yorker*, July 28, 2003.

7. Timothy Leary, *High Priest* (New York: New American Library, 1968), 183–84. See also Timothy Leary, *Flashbacks: An Autobiography* (New York: Putnam, 1983).

8. Ralph Metzner, "From Harvard to Zihuatenejo," in Robert Forte, ed., *Timothy Leary: Outside Looking In* (Rochester, Vt.: Park Street Press, 1999), 162.

9. Editorial, *New York Times*, February 16, 1929.

10. Aldous Huxley has the mystic Mr. Propter say this in his first American novel, *After Many a Summer Dies the Swan* (1939; reprinted Chicago: Elephant Paperbacks, 1993), 189.

11. Talcott Parsons to McGeorge Bundy recommending Bales's promotion, May 9, 1955, Harvard Archives, UAV 801.2010.

12. Conversation between Bales and Leary quoted in Patrick Schmidt, *Towards a History of the Department of Social Relations*, B. A. Honors Thesis, Harvard College (March 1978), 81.

13. Stanley Milgram, "Some Conditions of Obedience and Disobedience to Authority," *Human Relations* 18 (1965): 57.

14. Interview with an anonymous Milgram subject in Lauren Slater, *Opening Skinner's Box: Great Psychological Experiments of the Twentieth Century* (New York: W. W. Norton, 2004), 60.

15. Benjamin Harris, "Keywords: The History of Debriefing in Social Psychology," in Jill G. Morawski, ed., *The Rise of Experimentation in American Psychology* (New Haven, Conn.: Yale University Press, 1988), 196.

16. Susan Sontag, "Regarding the Torture of Others," *New York Times Magazine*, May 23, 2004, 27.

CONCLUSION

1. Herbert Marcuse, preface to *Eros and Civilization* (Boston: Beacon Press, 1966), xiv.

ACKNOWLEDGMENTS

Working on this book for the past several years has been like having a best friend who goes with me wherever I go and engages me in constant conversation. Why this is I'm not entirely sure, as the subject matter—loosely speaking, the development of psychological and social engineering techniques in America—is not necessarily a cozy, have-a-cup-of-tea one. Still, writing it has been a way for me to make sense of my own experience and, to some extent, others'. (I was trained as an anthropologist, and this is my version of anthropology.)

Lots of people have helped in this process—in particular, teachers. At Berkeley, Paul Rabinow inspired me through his Socratic teaching method and through his ability not to accept easy answers. David Hollinger's work made me want to write about American intellectual history. Alan Dundes helped get me started on this path and was continually supportive, for which I will always be grateful. Almost twenty years ago, Leslie Brisman of the English Department at Yale caught my attention with his reading of *Paradise Lost*—creative truth, he said, is found not *in* but *between* things.

For her generous reading of chapter 7, her helpful corrections, and her permission to quote from her husband's correspondence, I would like to thank Joan Dollard; I hope I have at least made a start at capturing the complexity and significance of John Dollard's work.

Many archivists—in particular Thomas Rosenbaum at the Rockefeller Archive Center and Milton Gustafson of the National Archives—were extremely helpful in pulling out sometimes obscure and rarely-asked-for documents. I received support for my research from the Regents of the University of California, the Rockefeller Archive Center, the Humanities Research Institute, and as a Woodrow Wilson Postdoctoral Fellow. To Kathy Woodward I owe many thanks for her encouragement and positive attitude during my time at the University of Washington.

I feel especially lucky to have had Kristy McGowan as my editor, as she displayed an ideal mix of intuition and common sense in editing this book. To Thomas LeBien, Hill and Wang's publisher, I owe many thanks for "getting" this project when I first showed it to him and generously encouraging it in many later versions. Gair Crutcher, Laura Haddad, Marybeth Hamilton, Doug Lemov, Celia Lowe, Tim Power, and Sonnet Retman were all inspiring interlocutors on this and related topics. I am deeply grateful to my parents, Penelope and Michael Lemov, who have given me many opportunities and always encouraged my education. To Palo Coleman, who helped me think about the "world as laboratory" and helped me think in general, and to our daughter, Ivy, this first book is dedicated.

INDEX